# Epidermal Growth Factor

# METHODS IN MOLECULAR BIOLOGY™

## John M. Walker, SERIES EDITOR

METHODS IN MOLECULAR BIOLOGY™

# Epidermal Growth Factor

*Methods and Protocols*

Edited by

## Tarun B. Patel

*Department of Pharmacology and Experimental Therapeutics*
*Loyola University Chicago*
*Maywood, IL*

## Paul J. Bertics

*Department of Biomolecular Chemistry*
*University of Wisconsin*
*Madison, WI*

HUMANA PRESS ✳ TOTOWA, NEW JERSEY

Cover illustration: Figure 1 from Chapter 5, "Regulator of Epidermal Growth Factor Signaling: *Sprouty*," by Esther Sook Miin Wong and Graeme R. Guy.

Production Editor: Tracy Catanese

Cover design by Patricia F. Cleary

For additional copies, pricing for bulk purchases, and/or information about other Humana titles, contact Humana at the above address or at any of the following numbers: Tel.: 973-256-1699; Fax: 973-256-8341; E-mail: orders@humanapr.com; or visit our Website: www.humanapress.com

Printed in the United States of America. 10 9 8 7 6 5 4 3 2 1

eISBN 1-59745-012-X

ISSN 1064-3745

Library of Congress Cataloging in Publication Data

Epidermal growth factor : methods and protocols / edited by Tarun B. Patel, Paul J. Bertics.
 p. ; cm. -- (Methods in molecular biology ; 327)
 Includes bibliographical references and index.
 ISBN 1-58829-421-8 (alk. paper)
 1. Epidermal growth factor--Laboratory manuals.
 [DNLM: 1. Epidermal Growth Factor--analysis. 2. Epidermal Growth Factor--physiology. 3. Receptor, Epidermal Growth Factor--analysis. 4. Receptor, Epidermal Growth Factor--physiology. WK 170 E638 2006] I. Patel, Tarun B., 1953- II. Bertics, Paul J. III. Series: Methods in molecular biology (Clifton, N.J.) ; v. 327.
 QP552.E59E47 2006
 612.7'91--dc22

                                                                    2005016687

# Preface

In compiling *Epidermal Growth Factor: Methods and Protocols*, we have tried to include a wide spectrum of methods used to study this very important developmental and therapeutic target. These methods range from the study of the purified EGF receptor to complex signaling and processing networks in intact cells. Because the EGF receptor system is becoming increasingly recognized for its involvement in multiple aggressive cancers, it is also a major focus of clinical and pharmacological analyses. With this in mind, we have also incorporated a chapter on the clinical and pharmacological considerations in cancer therapy. We hope that the contributions of many of the leading experts in the field will help advance the basic and clinical research on EGF.

We would like to take this opportunity to thank all of the authors for their contributions, without which this volume would not have been possible. We also thank Lorraine Grimsby for her administrative help in the compilation of this volume.

*Tarun B. Patel*
*Paul J. Bertics*

# Contents

# Contributors

ANDRE T. BARON • *Division of Hematology, Oncology, Kentucky School of Public Health, Markey Cancer Center, University of Kentucky, KY*

PAUL J. BERTICS • *Department of Biomolecular Chemistry, University of Wisconsin, Madison, WI*

CARL P. BLOBEL • *Arthritis and Tissue Degeneration Program, Hospital for Special Surgery, Weill Medical College of Cornell University, New York, NY*

DEEPTI CHATURVEDI • *Department of Pharmacology and Experimental Therapeutics, Loyola University Chicago, Maywood, IL*

VALERIE CHESNEAU • *Arthritis and Tissue Degeneration Program, Hospital for Special Surgery, Weill Medical College of Cornell University, New York, NY*

PRAKASH CHINNAIYAN • *Department of Human Oncology, University of Wisconsin, Madison, WI*

IVAN DIKIC • *Institute for Biochemistry II, Goethe University School of Medicine, Frankfurt, Germany*

FRANCIS EDWIN • *Department of Pharmacology and Experimental Therapeutics, Loyola University Chicago, Maywood, IL*

OLIVER M. FISCHER • *Pfizer Global Research and Development, Sandwich Laboratories, Sandwich, Kent, UK*

DAVID L. FULGHAM • *Department of Biomolecular Chemistry, University of Wisconsin, Madison, WI*

LINDA GRIFFITH • *Department of Material Sciences, Massachusetts Institute of Technology, Cambridge, MA*

ARTURO G. GUADARAMMA • *Department of Biomolecular Chemistry, University of Wisconsin, Madison, WI*

GRAEME R. GUY • *Institute of Molecular and Cell Biology, Proteos, Singapore*

PAUL M. HARARI • *Department of Human Oncology, University of Wisconsin, Madison, WI*

BRIAN HARMS • *Biological Engineering Division, Department of Chemical Engineering, and Center for Cancer Research, Massachusetts Institute of Technology, Cambridge, MA*

STEFAN HART • *Centre for Molecular Medicine, Agency for Science, Technology and Research (A\*STAR), Singapore*

KEISUKE HORIUCHI • *Arthritis and Tissue Degeneration Program, Hospital for Special Surgery, Weill Medical College of Cornell University, New York, NY*

LAURIE G. HUDSON • *Program in Toxicology and Pharmacology, College of Pharmacy, University of New Mexico Health Sciences Center, Albuquerque, NM*

AKIHIRO IWABU • *Department of Pathology, University of Pittsburgh, Pittsburgh, PA*

MENACHEM KATZ • *Department of Biological Regulation, The Weizmann Institute of Science, Rehovot, Israel*

LILY KOO • *Biological Engineering Division, Department of Chemical Engineering, Department of Material Science, and Center for Cancer Research, Massachusetts Institute of Technology, Cambridge, MA*

JACQUELINE M. LAFKY • *Endocrine Research Unit, Mayo Clinic College of Medicine, Rochester, MN*

DOUGLAS A. LAUFFENBURGER • *Department of Chemical Engineering, Biological Engineering Division, and Center for Cancer Research, Massachusetts Institute of Technology, Cambridge, MA*

NITA J. MAIHLE • *Departments of Ob/Gyn and Reproductive Sciences, Pathology and Pharmacology, Yale University School of Medicine, New Haven, CT*

ADRIANO MARCHESE • *Department of Pharmacology and Experimental Therapeutics, Loyola University Chicago, and Stritch School of Medicine, Maywood, IL*

YARON MOSESSON • *Department of Biological Regulation, The Weizmann Institute of Science, Rehovot, Israel*

TARUN B. PATEL • *Department of Pharmacology and Experimental Therapeutics, Loyola University Chicago, Maywood, IL*

CHIMERA RENE PEET • *Department of Biomolecular Chemistry, University of Wisconsin, WI*

UMUT SAHIN • *Arthritis and Tissue Degeneration Program, Hospital for Special Surgery, Weill Medical College of Cornell University, New York, NY*

MIRKO H. H. SCHMIDT • *Institute for Biochemistry II, Goethe University School of Medicine, Germany*

RAKESH SINGH • *Department of Pharmacology and Experimental Therapeutics, Loyola University Chicago, Maywood, IL*

KIRSTY SMITH • *Biological Engineering Division, Department of Chemical Engineering, and Center for Cancer Research, Massachusetts Institute of Technology, Cambridge, MA*

HUI SUN • *Department of Medicine and Pediatrics, University of Massachusetts, Worcester, MA*

AXEL ULLRICH • *Department of Molecular Biology, Max-Planck-Institute of Biochemistry, Martinsried, Germany*

ALAN WELLS • *Department of Pathology, University of Pittsburgh, Pittsburgh, PA*

GISELA WESKAMP • *Arthritis and Tissue Degeneration Program, Hospital for Special Surgery, Weill Medical College of Cornell University, New York, NY*

GREGORY J. WIEPZ • *Department of Biomolecular Chemistry, University of Wisconsin, WI*

ESTHER SOOK MIIN WONG • *Institute of Molecular and Cell Biology, Proteos, Singapore*

YOSEF YARDEN • *Department of Biological Regulation, The Weizmann Institute of Science, Rehovot, Israel*

REEMA ZEINELDIN • *Department of Chemical and Nuclear Engineering, University of New Mexico, Albuquerque, NM*

YUFANG ZHENG • *Arthritis and Tissue Degeneration Program, Hospital for Special Surgery, Weill Medical College of Cornell University, New York, NY*

# 1

# A Historical Perspective of the EGF Receptor and Related Systems

Francis Edwin, Gregory J. Wiepz, Rakesh Singh,
Chimera Rene Peet, Deepti Chaturvedi, Paul, J. Bertics,
and Tarun B. Patel

## Summary

Since the isolation of epidermal growth factor (EGF) from mouse submaxillary glands in the early 1950s by Cohen and coworkers, this growth factor has been shown to have various effects on numerous cellular systems. The biological and physiological role that EGF plays during development and in adult animals led to the identification of its receptor (EGFR) as well as the other members of the EGF family of growth factors and their receptors. In this chapter we provide a historical overview of the discovery of EGF, identification of the other members of EGF family, early studies on the actions of EGF, as well as the discovery and structural characterization of its receptor. Further, we have reviewed the transactivation of the EGFR by agonists for G protein-coupled receptors (GPCRs) and other extracellular stimuli unrelated to EGF-like ligands. Finally, an overview of the role of the EGFR family members in various diseases, including different forms of cancer, is provided.

**Key Words:** Growth factors; ligand; EGF; epiregulin; ampiregulin; EGF receptor; dimerization; kinase domain; transmembrane domain; EGFR transactivation; GPCR; GPCR agonists; HB-EGF; VEGF; receptor tyrosine kinase; tyrosine phosphorylation; Src kinase; protein kinase C; MAP kinase; LPA; angiotensin II; PYK2; ADAM; TGF-α; Jak2; integrins; growth hormones; migration; proliferation; ErbB receptors; mutations; disease; cancer cells.

## 1. Discovery of EGF

Rita Levi-Montalcini, a developmental biologist, pioneered the idea of growth factors with the discovery of nerve growth factor (NGF) in early 1950s *(1)*. In 1953, Dr. Stanley Cohen joined the Levi-Montalcini group and began

From: *Methods in Molecular Biology, vol. 327: Epidermal Growth Factor: Methods and Protocols*
Edited by: T. B. Patel and P. J. Bertics © Humana Press Inc., Totowa, NJ

working on NGF and the isolation of growth factors from salivary glands. During the course of these studies he observed early development of newborn mice treated with salivary gland extract. Treated mice showed eye opening and tooth eruption as early as 6–7 d, while in control animals the same process took 12–14 d *(2,3)*. These observations led Cohen to propose that salivary gland extracts contain growth-promoting substances responsible for precocious development of mice. Further research on the system led Cohen to isolate the substance responsible for growth acceleration, which he called epidermal growth factor (EGF) because of its ability to stimulate the proliferation of murine epidermal and corneal epithelial cells *(3)*.

Amino acid sequence analysis revealed EGF to be a small peptide of 53 amino acids with 6 cysteine residues *(4)*. The cysteine residues within EGF form three intramolecular disulfide linkages that are essential for biological activity of growth factor *(4,5)*. For their pioneering work on growth factor identification, Stanley Cohen and Rita Levi-Montalcini shared the 1986 Nobel Prize in Physiology/Medicine.

## 2. EGF-Related Factors

To date, at least 11 different EGF-like polypeptides have been reported, which together comprise the EGF family of growth factors. Each of these peptide shares at least one EGF-like domain comprising disulfide linkages *(4,6)*. The EGF family includes transforming growth factor (TGF)-α, amphiregulin (AR), heparin-binding EGF (HB-EGF), epiregulin (EPR), betacellulin (BTC), neuregulins 1–4 (NRG1-4), and teratocarcinoma-derived growth factor (Cripto-1) *(6)*. Recent reports also suggest that EGFR can be activated by viruses and viral proteins, including Epstein-Barr virus (EBV)-encoded latent membrane protein 1 (LMP1), human cytomegalovirus (HCMV)-encoded protein, respiratory syncytial virus (RSV)-encoded protein *(7)*, and smallpox growth factor, a protein produced by viriola virus *(8)*.

## 3. Early Studies on EGF Actions

Stanley Cohen's initial report on the precocious development of newborn mice by EGF triggered a flurry of research to uncover the biological and physiological role that EGF plays during development and in adult animals. Besides the salivary gland, detectable levels of EGF were also found in various tissue extracts and body fluids, including amniotic fluid, milk, saliva, gastric and duodenal contents, pancreatic juice, bile, and urine *(9)*. The discovery of EGF led to its characterization as an embryotrophic factor, i.e., it enhances mitogenesis, development, and implantation in different mammalian species *(10)*. In addition, the presence of EGF mRNA in embryonic tissues suggests its involvement in early embryonic development *(11)*. Further research has also implicated EGF in the morphogenesis of many organs, including teeth, brain, reproductive tracts, skin, gastrointestinal tracts, and in cardiovascular differentiation *(12–15)*.

EGF-immunized animal studies confirm most of the earlier reports of EGF involvement in embryonic development *(16,17)*. These studies showed that rats developed dry, thinner, and wrinkled skin, poorly developed hair follicle, poor organ development, and significant lower weight *(16,17)*. In addition to its role in embryogenesis, it is also clear that EGF is produced in many adult animal tissues *(18)*. In this regard, EGF has been implicated in epithelial cytoprotection, tissue trophism, cell survival regulation, cardiac function *(15)*, and accelerated wound healing in experimental animals *(19,20)*. EGF has also been reported to play an essential role in normal epithelial regeneration, which occurs regularly in vital organs such as the gastrointestinal, genitourinary, respiratory, and corneal epithelia *(21–23)*. These diverse actions are consistent with the identification of EGFR on cells of different origin, including fibroblast, corneal cells, and breast cells *(24)*. In this respect, because of its pleiotropic actions (*see* below), the term "epidermal growth factor" does not reflect its wide array of actions.

Over the past several decades, numerous reports have appeared describing the opposing actions of EGF on processes such as proliferation *(25,26)* and apoptosis *(27,28)* in many different cell types. Among other possibilities (*see* below), these effects have been postulated to be a feature of receptor expression levels, cell surface receptor density, and ligand concentration, all of which can lead to biphasic responses *(29)*.

## 4. Identification and Structural Aspects of the EGF Receptor

After its discovery, EGF was extensively studied for its mitogenic effects, which included accelerated hair growth, epidermal keratinization, stimulation of human skin fibroblast growth, and hypertrophy/hyperplasia of corneal epithelial cells in organ culture *(30,31)*. These and other studies prompted a search for cell surface receptor for EGF, and in the late 1970s, crosslinking of [125I]EGF to proteins on A431 cell membranes identified a protein of approximately 170 kDa as the putative EGFR *(32–34)*. Besides binding to EGF, this protein was ultimately identified to be a transmembrane glycoprotein that possesses an intrinsic tyrosine kinase activity stimulated by EGF binding *(34)*.

Cloning and sequencing studies revealed the EGFR to be a 1186-amino-acid protein with a predicted extracellular ligand binding domain, a single transmembrane domain, followed by an intracellular region that contains the tyrosine kinase domain and a regulatory domain in the carboxy terminus *(35)*. To date, three additional members of the EGFRs have been discovered (for review, *see* **ref. 6**). Thus, the four family members are ErbB1 (EGFR), ErbB2/

HER2/Neu, ErbB3/HER3, and ErbB4/HER4 *(6)*. Each of these receptors exhibits structural similarities in its extracellular, transmembrane, and cytosolic domains.

Until recently, other than its amino acid sequence, relatively little was known about the structure of the EGFR. However, in the last few years, a number of crystal structures have been solved for the extracellular region in both the ligand-bound and unbound states *(36,37)*. Likewise, the crystal structure of the intracellular tyrosine kinase domain *(38)* as well as the structural features of the other members of the EGF receptor family have been characterized and are reviewed elsewhere *(39,40)*. These models have added a new dimension to our understanding of how the EGFR interacts with its ligands and forms dimers with itself and other family members. Additionally, these models have also provided some insight into receptor activation and the mechanisms regulating the intracellular tyrosine kinase domain.

The extracellular region of the EGF receptor has been characterized as having four distinct subdomains (I–IV), which are composed of two sets of tandem repeats that primarily make up the ligand binding site. Two of the homologous regions (domains II and IV) are highly enriched in cysteine residues that form numerous disulfide bonds within each region *(41,42)*. In the absence of any ligand, the overall structure of the extracellular domain appears to be locked into an autoinhibitory configuration *(41)*. The purpose of this resting confirmation appears to be to prevent dimerization motifs from being exposed. Scatchard analyses indicated that the receptor possessed two different binding affinities: a low-affinity $K_d$ of ~0.1 n$M$ and high-affinity $K_d$ of ~1–10 n$M$ *(43)*. However, the mechanism regulating the different affinities has remained elusive. The unbound receptor resides predominantly in a low-affinity state *(43)*, but a fraction of the total population appears to exist in the high-affinity state associated with the insoluble cytoskeletal elements *(44)*. The recently defined crystal structures have helped to explain the mechanism behind the existence of the two affinity states. Initially, the ligand is believed to associate with only domain I in a low-affinity state (~10 n$M$) *(41)*. Once the initial ligand-receptor association is established, a conformational change occurs in the extracellular region due to rupturing of hydrogen bonds between domains II and IV (the cysteine-rich regions), thereby allowing ligand interaction with domain III, which results in a high affinity interaction *(41)*.

The aforementioned conformational change is believed to induce the formation of receptor dimers *(42)*. In addition, it is well known that upon binding of ligand, the EGFR will form homodimers with another EGFR *(45)* or heterodimers with another member of the EGFR family (ErbB2, ErbB3, or ErbB4) *(40)*. Recently described crystal structures predict the presence of a "dimerization arm" in the extracellular domain *(36,37)*. It has been proposed that when the ligand binds to the EGFR, there is a conformational change that exposes a loop from domain II, thereby permitting its interaction with another ligand-bound EGFR or ErbB family member to form the dimer. Additionally, there appear to be regions in domain IV that may also play a role in dimer formation and stabilization *(46,47)*. Interestingly, however, it has also been shown that domain IV appears to have a negative regulatory role in EGFR activation, because deletion of this region increases the overall activity of the receptor *(48)*.

The EGFR possesses a single transmembrane domain of 23 amino acids that anchors the receptor to the membrane but also appears to be involved in receptor dimerization *(39)*. The transmembrane domain also enhances the ability of dimers to form when attached to the extracellular portion *(49)*, and mutations in ErbB2 transmembrane domain have also been shown to augment its dimerization *(50)*. Additionally, when cells are exposed to peptides corresponding to the transmembrane domain of the EGFR, receptor autophosphorylation and downstream signaling events are inhibited *(51)*. It also appears that the transmembrane domain is involved in aligning the intracellular kinase domains upon dimerization through rotational twisting *(52,53)*.

The intracellular domain contains multiple domains, most notably the tyrosine kinase domain (~260 aa), a juxtamembrane region (~40 aa), and a carboxy-terminal regulatory region (~232 aa) *(54)*. This regulatory region of the receptor also contains numerous autophosphorylation sites that act as docking sites for SH2 domain proteins *(55)*. Recent evidence from the defined crystal structures suggests that the tyrosine kinase is in a constitutively active conformation but remains inaccessible due to the regulatory region of the carboxy terminus *(38)*, which is consistent with earlier kinetic studies *(56)*. It is believed that ligand binding exposes the dimerization arms and allows the receptors to come into close proximity such that the receptors can transphosphorylate tyrosine residues, undergo conformational changes, and remove the inhibitory constraint on the kinase domain *(57)*. The EGFR is now capable of phosphorylating tyrosine residues on other substrates such as phospholipase C$\gamma$ *(58)*, enolase, and various cytoskeletal-associated proteins such as $\alpha$-actinin *(59)* and protein 4.1 *(60)*.

In addition to the autophosphorylated tyrosines, there are residues that are phosphorylated by nonreceptor kinases including tyrosine residues *(61,62)*, as well as serine and threonine residues *(63,64)*. These sites are phosphorylated by downstream kinases that are part of the EGFR-activation cascade *(55)*. For instance, threonine 654 is a target for phosphorylation by protein kinase C *(65)*, and tyrosine 845 is a substrate for the tyrosine kinase Src *(66)*, whereas several serine residues can be phosphorylated by activated PKA *(67)*. Among the residues phosphorylated, the phospotyrosines can serve as docking sites for various proteins that contain Src homology 2 (SH2) or phosphotyrosine-binding domains (PTBs) *(55)* and thus serve as a recruitment center for members of various cell-signaling cascades.

## 5. Transactivation of the EGFR

Besides being activated by its cognate ligands, the EGFR can also acts as a signaling partner with other receptors outside its own family. Cross-communication between heterologous signaling systems and the EGFR has been shown to be critical for a variety of biological responses. Agonists for G protein coupled receptors (GPCRs) and other extracellular stimuli unrelated to EGF-like ligands, including agonists for cytokine receptor family members (prolactin, growth hormone), adhesion receptors (integrins), membrane-depolarizing agents (KCl), and environmental stress factors (ultraviolet and $\gamma$ irradiation, oxidants, heat shock, hyperosmotic shock) activate the EGF receptor in several cell systems. This phenomenon is referred to as inter-receptor cross-talk or EGFR signal transactivation *(68–71)*. Several studies indicate that the EGFR transactivation mechanism is subject to different cell type-specific regulatory influences.

### 5.1. Transactivation of the EGFR and Its Family Members by GPCRs

EGFR signal transactivation by GPCRs was originally described by Daub and colleagues *(68)*. This cross-talk mechanism has been further established in a variety of cell types such as human keratinocytes, primary mouse astrocytes, PC-12 cells, vascular smooth muscle cells, and cancer cells and considered as a widely relevant pathway towards the activation of the mitogen-activated protein (MAP) kinase signaling *(72–74)*. EGFR and HER2/neu are rapidly tyrosine phosphorylated after stimulation of Rat-1 cells with the GPCR agonists endothelin-1 (ET-1), lysophosphatidic acid (LPA), or thrombin *(68)*. Interestingly, LPA-induced transactivation of the EGFR in COS-7 cells was attenuated by pertussis toxin (PTX), which inactivates G$\alpha$ subunits of the G$_{i/o}$ family of G proteins. In contrast, thrombin-stimulated EGFR tyrosine phosphorylation and downstream signaling were not affected by PTX *(72)*. Moreover, lysophospholipids have recently been shown to transactivate HER2/neu in human gastric cancer cells *(75)*, while CXCR4 activation induces EGFR transactivation in an ovarian cancer cell line *(76)*. The transactivation of receptor tyrosine kinase couples GPCR activation to the ERK pathway, induction of *c-fos* gene expression, and DNA synthesis, which are abrogated either by the selective EGFR inhibitor tyrphostin AG1478 or by expression of a dominant-negative EGFR mutant *(68)*. In addition, studies in vascular smooth muscle cells suggested that the transactivation of EGFR is an obligatory event for stimulation of protein synthesis by GPCRs *(77)*.

It has been shown that $G\alpha_{13}$ subunits mediate LPA-induced actin polymerization and actin stress fiber formation in Swiss 3T3 cells and mouse fibroblasts via EGFR transactivation *(78,79)*. In PC-12, vascular smooth muscle cells and intestinal epithelial cells elevation in intracellular $Ca^{2+}$ concentrations have been demonstrated to be necessary in $G_q$-coupled receptor-mediated EGFR transactivation *(73,74,80–82)*. Activation of the Ser/Thr protein kinase C (PKC) has also shown to be required for $G_q$-coupled receptors to induce EGFR transactivation in cell lines such as HEK-293 and PC-12 cells *(81,83,84)*. Besides the function of PKC in GPCR-mediated EGFR transactivation, Matsubara and coworkers reported that in cardiac myocytes, angiotensin II transactivated the EGFR via a $Ca^{2+}$/calmodulin-dependent mechanism *(80)*. Similarly in PC-12 cells, Zwick and colleagues *(85)* demonstrated the involvement of a $Ca^{2+}$-calmodulin-dependent kinase II (CaMK II) activity in $K^+$-, but not bradykinin-induced EGFR transactivation (**Fig. 1**). Several reports also suggested a role for proline-rich tyrosine kinase 2 (PYK2) in $G_q$-mediated EGFR tyrosine phosphorylation in PC-12 *(81)* and intestinal epithelial cells *(86)*. However, bradykinin-mediated EGFR transactivation in PC-12 cells appears to be PYK2 independent, but $Ca^{2+}$ dependent *(85)*.

Some of the mechanisms involved in EGFR transactivation are being made apparent by more recent studies. The GPCR-EGFR cross-talk involves cell surface processing of proHB-EGF by metalloproteinases of the ADAM (a disintegrin and a metalloproteinase) family in response to GPCR stimulation *(87,88)*. GPCR ligands LPA, carbachol, and bombesin have been shown to induce the proteolytic processing of the transmembrane proHB-EGF precursor to yield the mature ligand. Blocking of this process with either the metalloprotease inhibitor batimastat or the HB-EGF antagonistic diphtheria toxin mutant CRM197 completely abrogated GPCR-induced EGFR transactivation and Shc tyrosine phosphorylation *(89)*. A variety of metalloproteinases of the ADAM and matrix metalloproteinase families have been implicated in HB-EGF precursor shedding (for review, *see* **ref. *90***), including ADAM10 *(91)* and MDC9/ADAM9 *(92,93)* and ADAM12 *(94)*, whereas ADAM17 has been implicated as the major convertase of epiregulin, TGF-$\alpha$, ampiregulin, and HB-EGF-like growth factor *(95)* (**Fig. 1**). The signaling molecules involved in between the GPCR and the ADAM proteins have not been identified.

In addition to the EGFR, other receptor tyrosine kinases (RTKs) have shown to be activated by GPCR ligands. For example, insulin-like growth factor receptor (IGF-1R) phosphorylation is induced by thrombin *(96)*, while type 2 vascular endothelial growth factor (VEGFR-2) in human umbilical vein endothelial cells (HUVECs) is transactivated by sphingosine 1-phosphate

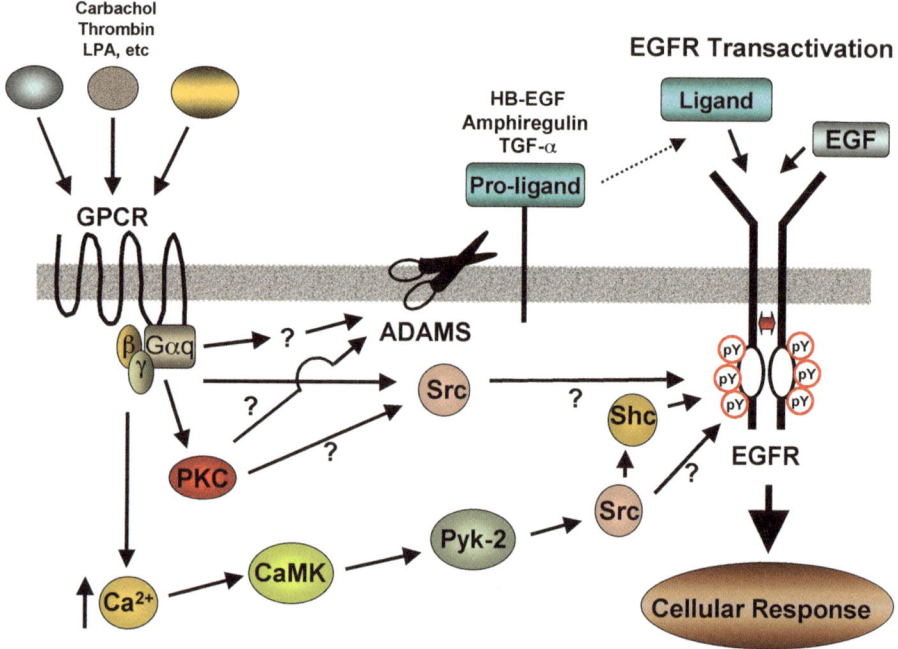

Fig. 1. Mechanisms of epidermal growth factor receptor (EGFR) transactivation. The EGFR is transactivated in a variety of cell systems through a number of different pathways, as shown. Some reports have provided evidence that elevation of intracellular $Ca^{2+}$ concentrations and activation of protein kinase C (PKC) are necessary for the G protein coupled receptor (GPCR)-mediated EGFR transactivation in different cell types. ADAMs have been identified as the major metalloproteinases that proteolytically process the pro-ligand precursors. Src has been shown to indirectly activate EGFR and has often been found associated with the activated EGFR. Details of several mechanisms involved in these processes that remain to be elucidated are shown by arrows with "?".

(S1P) (97). Moreover, LPA induces platelet-derived growth factor receptor (PDGFR) tyrosine phosphorylation in L cells (98), and opioid receptor agonists transactivate the fibroblast growth factor receptor-1 in rat C6 glioma cells that lack the EGFR (99). Furthermore, agonist stimulation of ectopically expressed $G_q$-coupled bombesin receptor or $G_i$-coupled M2 muscarinic acetylcholine receptor has been shown to trigger EGFR transactivation followed by tyrosine phosphorylation of Shc and formation of Shc-Grb2 complexes (100). Some of the receptor tyrosine kinases noted above may be phosphosphorylated following heterodimerization with the EGFR. For instance, the mechanism of EGFR transactivation by PDGF was shown to be due to the interaction between

PDGFR and EGFR through hetero-dimerization of the receptors *(101)*. Other mechanisms similar to the release of HB-EGF may exist for transactivation of receptor tyrosine kinases that do not heterodimerize with the EGFR. Additionally, depending upon the time frame of transactivation, signals emanating from GPCRs may stimulate the expression of growth factors that then activate their own receptors. Such a scenario is determined by the increase in VEGF synthesis in response to constitutive active Kaposi's sarcoma-associated herpesvirus (KSHV)-GPCR through the activation of transcription factor hypoxia-inducible factor-1α *(102)*. Similarly, angiotensin (Ang) II induces angiopoietin 2 and VEGF expression, thereby stimulating the processing of pro-HB-EGF by metalloproteinases, and the released HB-EGF transactivates EGFR to induce angiogenesis via the combined effect of angiopoietin 2 and VEGF *(103)*.

### 5.2. Transactivation of the EGFR by Activators of p60c-src

EGFR activation by GPCR agonists was assumed to exclusively rely on intracellular elements such as $Ca^{2+}$, PKC, and Src *(104)* (**Fig. 1**). Tyrosine phosphorylated Src is often found in association with the EGFR *(105)* or with PYK2 *(81,86)* upon stimulation of $G_q$-coupled receptors and therefore has been proposed to function as a mediator of EGFR transactivation. Though some reports have demonstrated Src-independent EGFR transactivation, Shc tyrosine phosphorylation, and ERK activation by Src, it seems likely that Src is recruited by the transactivated EGFR and thereby contributes to activation of the Ras signaling pathway *(72,106,107)*.

$H_2O_2$-induced transactivation of EGFR requires Src and mediates Erk1/2, but not Akt, activation in renal cells *(108)*. The muscarinic agonist carbachol (CCh) transactivates the EGFR via calmodulin, Pyk-2, and Src kinase activation *(86)*. CCh-stimulated EGFR transactivation and subsequent ERK activation, a process that limits CCh-induced chloride secretion, is mediated by metalloproteinase-dependent extracellular release of TGF-α and intracellular Src activation *(109)*.

### 5.3. Transactivation of the EGFR by Growth Hormone and Cytokines

When growth hormone binds to its receptor, which belongs to the cytokine receptor superfamily, it activates the Janus kinase Jak2. Jak2 provides tyrosine kinase activity and initiates activation of several key intracellular proteins (e.g., members of signal transduction and activator of transcription [STAT] family) that eventually execute the biological actions induced by growth hormone, including the expression of particular genes. In contrast to receptors that themselves have tyrosine kinase activity, the signaling pathways leading to MAP kinase activation triggered by growth hormone are poorly understood, but they appear to be mediated by the proteins Grb2 and Shc. Growth hormone stimulates tyrosine phosphorylation of EGFR and its association with Grb2 *(110)*. Studies using EGFR and its mutants have revealed that growth-hormone-induced activation of MAP kinase and expression of the transcription factor c-fos requires phosphorylation of tyrosines on EGFR, but not its own intrinsic tyrosine kinase activity *(110)*. The tyrosine at residue 1068 of the EGFR is proposed to be one of the important phosphorylation sites, and a Grb2-binding site stimulated by growth hormone via Jak2 *(111)*. These results indicate that the growth-hormone-stimulated phosphorylation of EGFR by Jak2 provides docking sites for Grb2 and activating MAP kinases and gene expression, independent of the intrinsic tyrosine kinase activity of EGFR. This may represent a novel cross-talk pathway between the cytokine receptor superfamily and growth factor receptor *(110)*.

### 5.4. Transactivation of the EGFR by Integrins

Like GPCRs, integrins have no intrinsic enzymatic activity, but couple to cytoplasmic proteins through short intracellular tails. A key player in integrin-mediated signaling is the focal adhesion kinase, a nonreceptor tyrosine kinase that interacts with docking or adapter proteins, including paxillin, tensin, and Grb2/SOS *(112,113)*. Integrin-dependent EGFR activation is a novel signaling mechanism involved in cell survival and proliferation in response to the extracellular matrix. Li et al. *(114)* reported that the integrin-mediated migration of murine B82L fibroblasts is dependent on the expression of intact EGFR, while EGF alone had little effect on the migration in the absence of extracellular matrix components. The molecular mechanisms underlying EGFR activation by integrins remain to be elucidated. Integrin-dependent activation of the EGFR and of other tyrosine kinases such as p125Fak occurs through distinct mechanisms *(115,116)*. Cytochalasin D, which disrupts the actin cytoskeleton, strongly inhibits p125Fak tyrosine phosphorylation, but does not affect EGFR tyrosine phosphorylation in response to integrins *(117)*. This indicates that the organization of actin cytoskeleton is not a primary event in integrin-dependent EGFR activation, while it is required to bring together signaling proteins leading to p125Fak tyrosine phosphorylation. Moreover, integrins and EGF receptors can associate on the membrane, forming a molecular complex, while association of integrins with p125Fak has not been detected *(118)*. Preliminary experiments also show that PtdIns-3 kinase is not involved in integrin-induced EGFR tyrosine phosphorylation, as detected by the use of specific inhibitors *(119)*.

### 6. Pleiotropic Actions of EGF

As noted above, EGF was first isolated from mouse submaxillary glands and was found to have an effect on premature eyelid and incisor eruption in newborn mice *(2,120)*. Since that time EGF has been shown to have various effects on numerous cellular systems *(121)*. The most commonly cited effect of EGF is its ability to induce mitogenesis in various cells types of ectodermal and mesodermal origin *(26,121)*. But EGF has also been shown to promote secretion of hormones such as prolactin *(122)* and human chorionic gonadotropin *(123)* as well as pituitary hormones and steroids *(121)*. EGF also induces the differentiation of specific cell types including the differentiation of human trophoblasts to syncytiotrophoblasts *(124)*. Additionally, EGF has been shown to influence glucose metabolism *(125)*.

One of the major factors that allow EGF to have such diverse and robust effects is that almost all cell types possess EGFRs with the exception of hematopoietic tissues *(121)*. The highest levels of EGF are found in the brain, thyroid, lung, liver, skin, placenta, and fetal membranes *(121)*. In addition, most tissues produce, contain and are exposed to sufficient levels of EGF or other ligands to activate the receptor *(121,126)*.

To add to the overall complexity of the system and as discussed above, the EGFR family is composed of four separate members (ErbB1, ErbB2, ErbB3, and ErbB4), of which there are at least 11 possible ligands (EGF, TGF-$\alpha$, epiregulin, amphiregulin, betacellulin, HB-EGF, Cripto-1, and neuregulins 1–4) that react with one or more of the receptors, which can then homo- or heterodimerize in different combinations *(40,127)*. This dimerization contributes to the activation of the various EGFR receptor family members and allows for the initiation of distinct biological processes *(39)*. Although many of the signaling pathways are similar between the different family members, their overall effects depend on the combination of ligands and receptor dimer composition *(127,128)*. One of the main effects appears to be on the kinetics involved in receptor downregulation. With the exception of the EGFR, the other family members are endocytosis impaired *(129)*, and formation of heterodimers with the EGFR can modify normal receptor downregulation.

The importance of the EGFR system is underscored by the observation that knockout mice lacking the EGFR die during gestation or shortly thereafter *(130–132)*. The EGFR-null mice that survive to birth are characterized by central nervous system (CNS) defects and extensive neurodegeneration *(131)*. Depending on their strain type, the EGFR-null mice that survive to birth display abnormalities in skin, kidney, brain, liver, and gastrointestinal tract *(132)*. Additionally, when knockouts of the EGFR ligands or combinations of ligands were evaluated, numerous cell systems were disrupted, including eye defects, epithelial cells of the skin, gastrointestinal tract, and tooth, and underdeveloped mammary glands *(133)*.

## 7. Role of EGFR Family Members in Disease

Recognizing the central role that ErbB receptors and their ligands play in proliferation, migration, cell survival, and transformation has made them an attractive target in the development of cancer therapies. ErbB receptors were first implicated in cancer when the avian erythroblastosis tumor virus (AEV) was found to carry v-ErbB, a mutant homolog of the human EGFR (ErbB1). This avian tumor virus was found to promote erythroleukemia and fibrosarcomas in chickens *(134,135)*. Since these original studies on v-ErbB, it has been realized that overexpression of ErbB receptors (and their ligands) as well as the expression of aberrant forms of these receptors can promote disease (cancer) progression *(136)*.

It has been shown that ErbB receptors play a pivotal role in cell transformation and progression of many carcinomas, including breast, ovarian, renal, non-small cell lung (NSCLC), head and neck, colorectal, pancreas, brain (glioma), bladder, esophagus, stomach, prostate, melanoma, thyroid, and endometrial cancers *(136)*. Overexpression of the EGFR, HER2, and ErbB ligands, such as EGF and TGF-α, is commonly observed in a wide range of diseases. For example, it is often found that tumor cells that overexpress ErbB receptors also produce ErbB-specific ligands, and this combination allows for cancer cells to activate themselves in an autocrine fashion. Overexpression of the EGFR alone (without a ligand) is not enough to promote disease, unless a constitutively active mutant receptor is expressed (*see* below). Although HER2 (ErbB2) overexpression has been largely associated with breast and ovarian cancers, it also plays a role in the development/progression of other cancers *(137)*. It is thought that under normal circumstances HER2 heterodimerization with other ErbB receptors is needed to activate downstream events *(138)*. Interestingly, several studies suggest that HER-2 overexpression or mutation is sufficient to promote cellular transformation *(138,139)*. ErbB3 and ErbB4 have also been found to be overexpressed in tumors, but not as much is known about their roles in disease *(140,141)*.

Even though misregulation and overexpression of ErbB receptors and their ligands is often associated with cancer, several lines of evidence suggest that alterations in ErbB status can also contribute to nonmalignant diseases. Most of the research on ErbB family members and their association with nonmalignant disease has focused on the participation of the EGFR and the ErbB1 ligands (EGF and TGF-α) in diseases such as polycystic kidney disease (PKD), cirrhosis, inflammatory bowel disease, and chronic asthma *(142,143)*.

Although overexpression of the ErbB receptors is thought to be the most common occurrence in ErbB driven diseases, mutant forms of the EGFR have also been found to be associated with tumorigenesis. For example, mutations in v-ErbB are thought to be important in the initiation and progression of cancers in chickens infected with the AEV *(134,135)*. As a retrovirus, AEV can promote cancer by insertionally activating a proto-oncogene (i.e., by integrating the retroviral genome next to proto-oncogenes) such as erbB1. In the case of AEV, the virus is believed to have originally integrated into the host genome within the erbB1 proto-oncogene. This insertion led to a truncation of the ErbB1-binding domain, resulting in the expression of a constitutively active receptor (v-ErbB), providing an advantage for viral replication and leading to development of leukemias in chickens. Subsequently, the viral gene accumulated mutations such as C-terminal truncations and point mutations that facilitated its ability to promote leukemias and sarcomas. Depending on the viral strain, the viral oncogene may carry one or more of the following mutations: a 267-amino-acid amino-terminal deletion that prevents ligand binding and causes constitutive tyrosine kinase activity, several substitution mutations mainly found in the kinase domain, a 21-amino-acid internal deletion, and a 74-amino-acid truncation from the cytoplasmic tail. Many of these mutations were characterized separately and have been found to be differentially capable of promoting leukemia and/or sarcoma. Specifically, the deletion of the ligand-binding domain, which creates a constitutively active tyrosine kinase domain, exhibits a high leukemogenic potential *(144)*.

One of the most frequently identified forms of aberrant human EGF receptor involved in cancer is EGFRvIII; this variant has a deletion of 270 amino acids at the N-terminus, which ablates part of the ligand-binding domain. Like v-ErbB, EGFRvIII is constitutively active and ligand independent. EGFRvIII is found in 40% of gliomas as well as a broad range of other carcinomas, such as medulloblastoma, prostate, breast, ovary, stomach, and lung (NSCLC) cancer *(136)*. In 2004, previously unidentified aberrant forms of EGF receptor were found in NSCLC patients. A subset of patients who responded well to gefitinib (a small-molecule tyrosine kinase inhibitor of the EGFR) had mutations in the kinase domain of their EGFR. The mutations in the kinase domain that conferred gefitinib sensitivity consisted of small in-frame 4- to 7-amino-acid deletions and several single amino acid substitutions *(8,145,146)*. These mutant EGFRs found in NSCLC still appear to be ligand dependent but promote higher proliferation rates in the presence of EGF *(147)*. Interestingly, these aberrant forms of EGFR found in NSCLC have not been detected in other cancers to date *(148)*.

Based on the information briefly reviewed above, it is clear that the EGFR and HER-2 are two ErbB family members that represent very promising molecular targets for disease therapy. Pharmacological inhibition of ErbB receptors appears ideal, because these systems are misregulated in many diseases and are important in growth and progression of disease (*see* Chapter 14). ErbB overexpression has been associated with a higher tumor grade, increased proliferation, and reduced survival of patients with many cancers *(136)*. The EGFR and associated ligands are also implicated in nonmalignant diseases, and for some of these disorders, such as PKD, targeting ErbB family members may prove beneficial *(143)*.

There are two general approaches to inhibit ErbB receptors currently used in cancer therapy: monoclonal antibodies (MAbs) directed against the ectodomain of receptors and small molecule tyrosine kinase inhibitors (TKIs) that target the catalytic tyrosine kinase domain. The MAbs were first designed in the 1980s against EGF receptor. Some MAbs used in clinical trials include cetuximab, directed against the EGFR (brand name Erbitux™), trastuzumab (directed against HER-2 (brand name Herceptin™), and MAb 806, which targets EGFRvIII (*see* also Chapter 14). In addition to the therapeutic utility of the ErbB-directed MAbs, the use of TKIs has also shown promise, and two small-molecule inhibitors of EGFR kinase activity being tested in clinical trials are gefitinib (Iressa™) and erlotinib (Tarceva™). As mentioned, recent studies have identified a cohort of patients with somatic mutations in the EGFR tyrosine kinase domain that appear to confer gefitinib sensitivity *(8,145,146)*. In vitro studies using cell lines harboring these kinase domain mutations, which include cells isolated from lung cancer patients as well as stable cell lines expressing aberrant forms of the receptor, have revealed that these cells are more sensitive to gefitinib than normal cells lacking the mutant receptors *(147)*. The discovery of tumor-specific receptors such as EGFRvIII and the gefitinib-sensitive mutant EGFRs should prove extremely valuable as therapeutic targets because these targeted therapies have the potential to be more selective. Unfortunately, most of the cancer treatments currently available are not very discriminating, and patients suffer side effects from broad-based therapies that target healthy and diseased cells alike. Conceivably, future research will find more disease-specific ErbB family variants that can be more selectively targeted to achieve more effective therapies.

## Acknowledgments

This work was supported by grants from the National Institutes of Health: CA105730, CA108467, and GM53271 to PJB and HL48308 and HL59679 to TBP.

## References

1. Levi-Montalcini, R., Meyer, H., and Hamburger, V. (1954) In vitro experiments on the effects of mouse sarcomas 180 and 37 on the spinal and sympathetic ganglia of the chick embryo. *Cancer Res.* **14,** 49–57.
2. Levi-Montalcini, R., and Cohen, S. (1960) Effects of the extract of the mouse submaxillary salivary glands on the sympathetic system of mammals. *Ann. NY Acad. Sci.* **85,** 324–341.

3. Cohen, S. (1965) The stimulation of epidermal proliferation by a specific protein (EGF). *Dev. Biol.* **12**, 394–407.
4. Savage, C. R., Jr., Inagami, T., and Cohen, S. (1972) The primary structure of epidermal growth factor. *J. Biol. Chem.* **247**, 7612–7621.
5. Savage, C. R., Jr., Hash, J. H., and Cohen, S. (1973) Epidermal growth factor. Location of disulfide bonds. *J. Biol. Chem.* **248**, 7669–7672.
6. Salomon, D. S., Brandt, R., Ciardiello, F., and Normanno, N. (1995) Epidermal growth factor-related peptides and their receptors in human malignancies. *Crit. Rev. Oncol. Hematol.* **19**, 183–232.
7. Monick, M. M., Cameron, K., Staber, J., et al. (2005) Activation of the epidermal growth factor receptor by respiratory syncytial virus results in increased inflammation and delayed apoptosis. *J. Biol. Chem.* **280**, 2147–2158.
8. Lynch, T. J., Bell, D. W., Sordella, R., et al. (2004) Activating mutations in the epidermal growth factor receptor underlying responsiveness of non-small-cell lung cancer to gefitinib. *N. Engl. J. Med.* **350**, 2129–2139.
9. St. Hilaire, R. J., Hradek, G. T., and Jones, A. L. (1983) Hepatic sequestration and biliary secretion of epidermal growth factor: evidence for a high-capacity uptake system. *Proc. Natl. Acad. Sci. USA* **80**, 3797–3801.
10. Chia, C. M., Winston, R. M., and Handyside, A. H. (1995) EGF, TGF-alpha and EGFR expression in human preimplantation embryos. *Development* **121**, 299–307.
11. Maruo, T., Matsuo, H., Otani, T., and Mochizuki, M. (1995) Role of epidermal growth factor (EGF) and its receptor in the development of the human placenta. *Reprod. Fertil. Dev.* **7**, 1465–1470.
12. Thesleff, I., Vaahtokari, A., and Partanen, A. M. (1995) Regulation of organogenesis. Common molecular mechanisms regulating the development of teeth and other organs. *Int. J. Dev. Biol.* **39**, 35–50.
13. Kato, M., Mizuguchi, M., and Takashima, S. (1995) Developmental changes of epidermal growth factor-like immunoreactivity in the human fetal brain. *J. Neurosci. Res.* **42**, 486–492.
14. Gupta, C. (1996) The role of epidermal growth factor receptor (EGFR) in male reproductive tract differentiation: stimulation of EGFR expression and inhibition of Wolffian duct differentiation with anti-EGFR antibody. *Endocrinology* **137**, 905–910.
15. Goldman, B., Mach, A., and Wurzel, J. (1996) Epidermal growth factor promotes a cardiomyoblastic phenotype in human fetal cardiac myocytes. *Exp. Cell Res.* **228**, 237–245.
16. Raaberg, L., Nexo, E., Jorgensen, P. E., Poulsen, S. S., and Jakab, M. (1995) Fetal effects of epidermal growth factor deficiency induced in rats by autoantibodies against epidermal growth factor. *Pediatr. Res.* **37**, 175–181.
17. Tsutsumi, O. and Oka, T. (1987) Epidermal growth factor deficiency during pregnancy causes abortion in mice. *Am. J. Obstet. Gynecol.* **156**, 241–244.
18. Kasselberg, A. G., Orth, D. N., Gray, M. E., and Stahlman, M. T. (1985) Immunocytochemical localization of human epidermal growth factor/urogastrone in several human tissues. *J. Histochem. Cytochem.* **33**, 315–322.

19. Playford, R. J., Wright, N. A., and Goodlad, R. A. (1995) Luminal nutrition and gut growth. *Gut* **36**, 943.
20. Hutson, J. M., Niall, M., Evans, D., and Fowler, R. (1979) Effect of salivary glands on wound contraction in mice. *Nature* **279**, 793–795.
21. Barrow, R. E., Wang, C. Z., Evans, M. J., and Herndon, D. N. (1993) Growth factors accelerate epithelial repair in sheep trachea. *Lung* **171**, 335–344.
22. Liu, A., Davis, R. J., Flores, C., Menon, M., and Seethalakshmi, L. (1992) Epidermal growth factor: receptor binding and effects on the sex accessory organs of sexually mature male mice. *J. Urol.* **148**, 427–431.
23. Stahlman, M. T., Orth, D. N., and Gray, M. E. (1989) Immunocytochemical localization of epidermal growth factor in the developing human respiratory system and in acute and chronic lung disease in the neonate. *Lab. Invest.* **60**, 539–547.
24. el-Galley, R. E., Smith, E., Cohen, C., Petros, J. A., Woodard, J., and Galloway, N. T. (1997) Epidermal growth factor (EGF) and EGF receptor in hypospadias. *Br. J. Urol.* **79**, 116–119.
25. Ullrich, A. and Schlessinger, J. (1990) Signal transduction by receptors with tyrosine kinase activity. *Cell* **61**, 203–212.
26. Carpenter, G. and Cohen, S. (1990) Epidermal growth factor. *J. Biol. Chem.* **265**, 7709–7712.
27. Cao, L., Yao, Y., Lee, V., Kiani, C., Spaner, D., Lin, Z., Zhang, Y., Adams, M. E., and Yang, B. B. (2000) Epidermal growth factor induces cell cycle arrest and apoptosis of squamous carcinoma cells through reduction of cell adhesion. *J. Cell. Biochem.* **77**, 569–583.
28. Armstrong, D. K., Kaufmann, S. H., Ottaviano, Y. L., et al. (1994) Epidermal growth factor-mediated apoptosis of MDA-MB-468 human breast cancer cells. *Cancer Res.* **54**, 5280–5283.
29. Kamer, A. R., Sacks, P. G., Vladutiu, A., and Liebow, C. (2004) EGF mediates multiple signals: dependence on the conditions. *Int. J. Mol. Med.* **13**, 143–147.
30. Cohen, S. and Elliott, G. A. (1963) The stimulation of epidermal keratinization by a protein isolated from the submaxillary gland of the mouse. *J. Invest. Dermatol.* **40**, 1–5.
31. Cohen, S. and Carpenter, G. (1975) Human epidermal growth factor: isolation and chemical and biological properties. *Proc. Natl. Acad. Sci. USA* **72**, 1317–1321.
32. Carpenter, G., Lembach, K. J., Morrison, M. M., and Cohen, S. (1975) Characterization of the binding of 125-I-labeled epidermal growth factor to human fibroblasts. *J. Biol. Chem.* **250**, 4297–4304.
33. Cohen, S., Carpenter, G., and King, L., Jr. (1980) Epidermal growth factor-receptor-protein kinase interactions. Co-purification of receptor and epidermal growth factor-enhanced phosphorylation activity. *J. Biol. Chem.* **255**, 4834–4842.
34. Cohen, S., Ushiro, H., Stoscheck, C., and Chinkers, M. (1982) A native 170,000 epidermal growth factor receptor-kinase complex from shed plasma membrane vesicles. *J. Biol. Chem.* **257**, 1523–1531.
35. Ullrich, A., Coussens, L., Hayflick, J. S., et al. (1984) Human epidermal growth factor receptor cDNA sequence and aberrant expression of the amplified gene in A431 epidermoid carcinoma cells. *Nature* **309**, 418–425.

36. Ogiso, H., Ishitani, R., Nureki, O., et al. (2002) Crystal structure of the complex of human epidermal growth factor and receptor extracellular domains. *Cell* **110,** 775–787.
37. Garrett, T. P., McKern, N. M., Lou, M., et al. (2002) Crystal structure of a truncated epidermal growth factor receptor extracellular domain bound to transforming growth factor alpha. *Cell* **110,** 763–773.
38. Stamos, J., Sliwkowski, M. X., and Eigenbrot, C. (2002) Structure of the epidermal growth factor receptor kinase domain alone and in complex with a 4-anilinoquinazoline inhibitor. *J. Biol. Chem.* **277,** 46,265–46,272.
39. Schlessinger, J. (2002) Ligand-induced, receptor-mediated dimerization and activation of EGF receptor. *Cell* **110,** 669–672.
40. Leahy, D. J. (2004) Structure and function of the epidermal growth factor (EGF/ErbB) family of receptors. *Adv. Protein Chem.* **68,** 1–27.
41. Ferguson, K. M., Berger, M. B., Mendrola, J. M., Cho, H. S., Leahy, D. J., and Lemmon, M. A. (2003) EGF activates its receptor by removing interactions that autoinhibit ectodomain dimerization. *Mol Cell* **11,** 507–517.
42. Burgess, A. W., Cho, H. S., Eigenbrot, C., et al. (2003) An open-and-shut case? Recent insights into the activation of EGF/ErbB receptors. *Mol. Cell.* **12,** 541–552.
43. King, A. C. and Cuatrecasas, P. (1982) Resolution of high and low affinity epidermal growth factor receptors. Inhibition of high affinity component by low temperature, cycloheximide, and phorbol esters. *J. Biol. Chem.* **257,** 3053–3060.
44. Gronowski, A. M. and Bertics, P. J. (1995) Modulation of epidermal growth factor receptor interaction with the detergent-insoluble cytoskeleton and its effects on receptor tyrosine kinase activity. *Endocrinology* **136,** 2198–2205.
45. Lax, I., Bellot, F., Howk, R., Ullrich, A., Givol, D., and Schlessinger, J. (1989) Functional analysis of the ligand binding site of EGF-receptor utilizing chimeric chicken/human receptor molecules. *EMBO J.* **8,** 421–427.
46. Saxon, M. L. and Lee, D. C. (1999) Mutagenesis reveals a role for epidermal growth factor receptor extracellular subdomain IV in ligand binding. *J. Biol. Chem.* **274,** 28,356–28,362.
47. Berezov, A., Chen, J., Liu, Q., Zhang, H. T., Greene, M. I., and Murali, R. (2002) Disabling receptor ensembles with rationally designed interface peptidomimetics. *J. Biol. Chem.* **277,** 28,330–28,39.
48. Elleman, T. C., Domagala, T., McKern, N. M., et al. (2001) Identification of a determinant of epidermal growth factor receptor ligand-binding specificity using a truncated, high-affinity form of the ectodomain. *Biochemistry* **40,** 8930–8939.
49. Tanner, K. G. and Kyte, J. (1999) Dimerization of the extracellular domain of the receptor for epidermal growth factor containing the membrane-spanning segment in response to treatment with epidermal growth factor. *J. Biol. Chem.* **274,** 35985–35,990.
50. Bargmann, C. I., Hung, M. C., and Weinberg, R. A. (1986) Multiple independent activations of the neu oncogene by a point mutation altering the transmembrane domain of p185. *Cell* **45,** 649–657.
51. Bennasroune, A., Fickova, M., Gardin, A., et al. (2004) Transmembrane peptides as inhibitors of ErbB receptor signaling. *Mol. Biol. Cell.* **15,** 3464–3474.

52. Moriki, T., Maruyama, H., and Maruyama, I. N. (2001) Activation of preformed EGF receptor dimers by ligand-induced rotation of the transmembrane domain. *J. Mol. Biol.* **311,** 1011–1026.
53. Bell, C. A., Tynan, J. A., Hart, K. C., Meyer, A. N., Robertson, S. C., and Donoghue, D. J. (2000) Rotational coupling of the transmembrane and kinase domains of the Neu receptor tyrosine kinase. *Mol. Biol. Cell* **11,** 3589–3599.
54. Walton, G. M., Chen, W. S., Rosenfeld, M. G., and Gill, G. N. (1990) Analysis of deletions of the carboxyl terminus of the epidermal growth factor receptor reveals self-phosphorylation at tyrosine 992 and enhanced in vivo tyrosine phosphorylation of cell substrates. *J. Biol. Chem.* **265,** 1750–1754.
55. Schlessinger, J. (2000) Cell signaling by receptor tyrosine kinases[see comment]. *Cell* **103,** 211–225.
56. Weber, W., Bertics, P. J., and Gill, G. N. (1984) Immunoaffinity purification of the epidermal growth factor receptor. Stoichiometry of binding and kinetics of self-phosphorylation. *J. Biol. Chem.* **259,** 14631–14,636.
57. Ferguson, K. M. (2004) Active and inactive conformations of the epidermal growth factor receptor. *Biochem. Soc. Trans.* **32,** 742–745.
58. Nishibe, S. and Carpenter, G. (1990) Tyrosine phosphorylation and the regulation of cell growth: growth factor-stimulated tyrosine phosphorylation of phospholipase C. *Sem. Cancer Biol.* **1,** 285–292.
59. Akiyama, T., Kadowaki, T., Nishida, E., et al. (1986) Substrate specificities of tyrosine-specific protein kinases toward cytoskeletal proteins in vitro. *J. Biol. Chem.* **261,** 14,797–14,7803.
60. Subrahmanyam, G., Bertics, P. J., and Anderson, R. A. (1991) Phosphorylation of protein 4.1 on tyrosine-418 modulates its function in vitro. *Proc. Natl. Acad. Sci. USA* **88,** 5222–5226.
61. Downward, J., Waterfield, M. D., and Parker, P. J. (1985) Autophosphorylation and protein kinase C phosphorylation of the epidermal growth factor receptor. Effect on tyrosine kinase activity and ligand binding affinity. *J. Biol. Chem.* **260,** 14,538–14,546.
62. Downward, J., Parker, P., and Waterfield, M. D. (1984) Autophosphorylation sites on the epidermal growth factor receptor. *Nature* **311,** 483–485.
63. Heisermann, G. J. and Gill, G. N. (1988) Epidermal growth factor receptor threonine and serine residues phosphorylated in vivo. *J. Biol. Chem.* **263,** 13,152–13,158.
64. Cochet, C., Gill, G. N., Meisenhelder, J., Cooper, J. A., and Hunter, T. (1984) C-kinase phosphorylates the epidermal growth factor receptor and reduces its epidermal growth factor-stimulated tyrosine protein kinase activity. *J. Biol. Chem.* **259,** 2553–2558.
65. Hunter, T., Ling, N., and Cooper, J. A. (1984) Protein kinase C phosphorylation of the EGF receptor at a threonine residue close to the cytoplasmic face of the plasma membrane. *Nature* **311,** 480–483.
66. Sato, K., Sato, A., Aoto, M., and Fukami, Y. (1995) c-Src phosphorylates epidermal growth factor receptor on tyrosine 845. *Biochem. Biophys. Res. Commun.* **215,** 1078–1087.

67. Barbier, A. J., Poppleton, H. M., Yigzaw, Y., et al. (1999) Transmodulation of epidermal growth factor receptor function by cyclic AMP-dependent protein kinase. *J. Biol. Chem.* **274,** 14,067–14,073.

68. Daub, H., Weiss, F. U., Wallasch, C., and Ullrich, A. (1996) Role of transactivation of the EGF receptor in signalling by G-protein-coupled receptors. *Nature* **379,** 557–560.

69. Luttrell, L. M., Daaka, Y., and Lefkowitz, R. J. (1999) Regulation of tyrosine kinase cascades by G-protein-coupled receptors. *Curr. Opin. Cell. Biol.* **11,** 177–183.

70. Zwick, E., Hackel, P. O., Prenzel, N., and Ullrich, A. (1999) The EGF receptor as central transducer of heterologous signalling systems. *Trends Pharmacol. Sci.* **20,** 408–412.

71. Marinissen, M. J. and Gutkind, J. S. (2001) G-protein-coupled receptors and signaling networks: emerging paradigms. *Trends Pharmacol. Sci.* **22,** 368–376.

72. Daub, H., Wallasch, C., Lankenau, A., Herrlich, A., and Ullrich, A. (1997) Signal characteristics of G protein-transactivated EGF receptor. *EMBO J.* **16,** 7032–7044.

73. Zwick, E., Daub, H., Aoki, N., et al. (1997) Critical role of calcium- dependent epidermal growth factor receptor transactivation in PC12 cell membrane depolarization and bradykinin signaling. *J. Biol. Chem.* **272,** 24,767–24,770.

74. Eguchi, S., Numaguchi, K., Iwasaki, H., et al. (1998) Calcium-dependent epidermal growth factor receptor transactivation mediates the angiotensin II-induced mitogen-activated protein kinase activation in vascular smooth muscle cells. *J. Biol. Chem.* **273,** 8890–8896.

75. Shida, D., Kitayama, J., Yamaguchi, H., et al. (2005) Lysophospholipids transactivate HER2/neu (erbB-2) in human gastric cancer cells. *Biochem. Biophys. Res. Commun.* **327,** 907–914.

76. Porcile, C., Bajetto, A., Barbero, S., Pirani, P., and Schettini, G. (2004) CXCR4 activation induces epidermal growth factor receptor transactivation in an ovarian cancer cell line. *Ann. NY Acad. Sci.* **1030,** 162–169.

77. Voisin, L., Foisy, S., Giasson, E., Lambert, C., Moreau, P., and Meloche, S. (2002) EGF receptor transactivation is obligatory for protein synthesis stimulation by G protein-coupled receptors. *Am. J. Physiol. Cell. Physiol.* **283,** C446–455.

78. Gohla, A., Harhammer, R., and Schultz, G. (1998) The G-protein G13 but not G12 mediates signaling from lysophosphatidic acid receptor via epidermal growth factor receptor to Rho. *J. Biol. Chem.* **273,** 4653–4659.

79. Gohla, A., Offermanns, S., Wilkie, T. M., and Schultz, G. (1999) Differential involvement of Galpha12 and Galpha13 in receptor-mediated stress fiber formation. *J. Biol. Chem.* **274,** 17,901–17,907.

80. Murasawa, S., Mori, Y., Nozawa, Y., et al. (1998) Role of calcium-sensitive tyrosine kinase Pyk2/CAKbeta/RAFTK in angiotensin II induced Ras/ERK signaling. *Hypertension* **32,** 668–675.

81. Soltoff, S. P. (1998) Related adhesion focal tyrosine kinase and the epidermal growth factor receptor mediate the stimulation of mitogen-activated protein kinase by the G-protein-coupled P2Y2 receptor. Phorbol ester or [Ca$^{2+}$]i elevation can substitute for receptor activation. *J. Biol. Chem.* **273,** 23,110–23,117.

82. Iwasaki, H., Eguchi, S., Ueno, H., Marumo, F., and Hirata, Y. (1999) Endothelin-mediated vascular growth requires p42/p44 mitogen-activated protein kinase and p70 S6 kinase cascades via transactivation of epidermal growth factor receptor. *Endocrinology* **140,** 4659–4668.

83. Tsai, W., Morielli, A. D., and Peralta, E. G. (1997) The m1 muscarinic acetylcholine receptor transactivates the EGF receptor to modulate ion channel activity. *Embo. J.* **16,** 4597–4605.

84. Grosse, R., Roelle, S., Herrlich, A., Hohn, J., and Gudermann, T. (2000) Epidermal growth factor receptor tyrosine kinase mediates Ras activation by gonadotropin-releasing hormone. *J. Biol. Chem.* **275,** 12,251–12,260.

85. Zwick, E., Wallasch, C., Daub, H., and Ullrich, A. (1999) Distinct calcium-dependent pathways of epidermal growth factor receptor transactivation and PYK2 tyrosine phosphorylation in PC12 cells. *J. Biol. Chem.* **274,** 20,989–20,996.

86. Keely, S. J., Calandrella, S. O., and Barrett, K. E. (2000) Carbachol-stimulated transactivation of epidermal growth factor receptor and mitogen-activated protein kinase in T(84) cells is mediated by intracellular ca(2+), PYK-2, and p60(src). *J. Biol. Chem.* **275,** 12,619–12,625.

87. Prenzel, N., Zwick, E., Daub, H., et al. (1999) EGF receptor transactivation by G-protein-coupled receptors requires metalloproteinase cleavage of proHB-EGF. *Nature* **402,** 884–888.

88. Fischer, O. M., Hart, S., Gschwind, A., and Ullrich, A. (2003) EGFR signal transactivation in cancer cells. *Biochem. Soc. Trans.* **31,** 1203–1208.

89. Gschwind, A., Zwick, E., Prenzel, N., Leserer, M., and Ullrich, A. (2001) Cell communication networks: epidermal growth factor receptor transactivation as the paradigm for interreceptor signal transmission. *Oncogene* **20,** 1594–600.

90. Blobel, C. P. (2005) ADAMs: key components in EGFR signalling and development. *Nat. Rev. Mol. Cell. Biol.* **6,** 32–43.

91. Yan, Y., Shirakabe, K., and Werb, Z. (2002) The metalloprotease Kuzbanian (ADAM10) mediates the transactivation of EGF receptor by G protein-coupled receptors. *J. Cell. Biol.* **158,** 221–226.

92. Roelle, S., Grosse, R., Aigner, A., Krell, H. W., Czubayko, F., and Gudermann, T. (2003) Matrix metalloproteinases 2 and 9 mediate epidermal growth factor receptor transactivation by gonadotropin-releasing hormone. *J. Biol. Chem.* **278,** 47,307–47,318.

93. Izumi, Y., Hirata, M., Hasuwa, H., et al. (1998) A metalloprotease-disintegrin, MDC9/meltrin-gamma/ADAM9 and PKCdelta are involved in TPA-induced ectodomain shedding of membrane-anchored heparin-binding EGF-like growth factor. *EMBO J.* **17,** 7260–7272.

94. Asakura, M., Kitakaze, M., Takashima, S., et al. (2002) Cardiac hypertrophy is inhibited by antagonism of ADAM12 processing of HB-EGF: metalloproteinase inhibitors as a new therapy. *Nat. Med.* **8,** 35–40.

95. Sahin, U., Weskamp, G., Kelly, K., et al. (2004) Distinct roles for ADAM10 and ADAM17 in ectodomain shedding of six EGFR ligands. *J. Cell. Biol.* **164,** 769–779.

96. Weiss, F. U., Daub, H., and Ullrich, A. (1997) Novel mechanisms of RTK signal generation. *Curr. Opin. Genet. Dev.* **7,** 80–86.
97. Endo, A., Nagashima, K., Kurose, H., Mochizuki, S., Matsuda, M., and Mochizuki, N. (2002) Sphingosine 1-phosphate induces membrane ruffling and increases motility of human umbilical vein endothelial cells via vascular endothelial growth factor receptor and CrkII. *J. Biol. Chem.* **277,** 23,747–23,754.
98. Herrlich, A., Daub, H., Knebel, A., et al. (1998) Ligand-independent activation of platelet-derived growth factor receptor is a necessary intermediate in lysophosphatidic, acid-stimulated mitogenic activity in L cells. *Proc. Natl. Acad. Sci. USA* **95,** 8985–8990.
99. Belcheva, M. M., Haas, P. D., Tan, Y., Heaton, V. M., and Coscia, C. J. (2002) The fibroblast growth factor receptor is at the site of convergence between mu-opioid receptor and growth factor signaling pathways in rat C6 glioma cells. *J. Pharmacol. Exp. Ther.* **303,** 909–918.
100. Won, S., Si, J., Colledge, M., Ravichandran, K. S., Froehner, S. C., and Mei, L. (1999) Neuregulin-increased expression of acetylcholine receptor epsilon-subunit gene requires ErbB interaction with Shc. *J. Neurochem.* **73,** 2358–2368.
101. Saito, Y., Haendeler, J., Hojo, Y., Yamamoto, K., and Berk, B. C. (2001) Receptor heterodimerization: essential mechanism for platelet-derived growth factor-induced epidermal growth factor receptor transactivation. *Mol. Cell. Biol.* **21,** 6387–6394.
102. Sodhi, A., Montaner, S., Patel, V., Zohar, M., Bais, C., Mesri, E. A., and Gutkind, J. S. (2000) The Kaposi's sarcoma-associated herpes virus G protein-coupled receptor up-regulates vascular endothelial growth factor expression and secretion through mitogen-activated protein kinase and p38 pathways acting on hypoxia-inducible factor 1alpha. *Cancer Res.* **60,** 4873–4880.
103. Fujiyama, S., Matsubara, H., Nozawa, Y., et al. (2001) Angiotensin AT(1) and AT(2) receptors differentially regulate angiopoietin-2 and vascular endothelial growth factor expression and angiogenesis by modulating heparin binding-epidermal growth factor (EGF)-mediated EGF receptor transactivation. *Circ. Res.* **88,** 22–29.
104. Carpenter, G. (1999) Employment of the epidermal growth factor receptor in growth factor-independent signaling pathways. *J. Cell. Biol.* **146,** 697–702.
105. Luttrell, L. M., Ferguson, S. S., Daaka, Y., et al. (1999) Beta-arrestin-dependent formation of beta2 adrenergic receptor-Src protein kinase complexes. *Science* **283,** 655–661.
106. Adomeit, A., Graness, A., Gross, S., Seedorf, K., Wetzker, R., and Liebmann, C. (1999) Bradykinin B(2) receptor-mediated mitogen-activated protein kinase activation in COS-7 cells requires dual signaling via both protein kinase C pathway and epidermal growth factor receptor transactivation. *Mol. Cell. Biol.* **19,** 5289–5297.
107. Slack, B. E. (2000) The m3 muscarinic acetylcholine receptor is coupled to mitogen-activated protein kinase via protein kinase C and epidermal growth factor receptor kinase. *Biochem. J.* **348 Pt 2,** 381–387.

108. Zhuang, S. and Schnellmann, R. G. (2004) H2O2-induced transactivation of EGF receptor requires Src and mediates ERK1/2, but not Akt, activation in renal cells. *Am. J. Physiol. Renal Physiol.* **286**, F858–865.

109. McCole, D. F., Keely, S. J., Coffey, R. J., and Barrett, K. E. (2002) Transactivation of the epidermal growth factor receptor in colonic epithelial cells by carbachol requires extracellular release of transforming growth factor-alpha. *J. Biol. Chem.* **277**, 42,603–42,612.

110. Yamauchi, T., Ueki, K., Tobe, K., et al. (1997) Tyrosine phosphorylation of the EGF receptor by the kinase Jak2 is induced by growth hormone. *Nature* **390**, 91–96.

111. Yamauchi, J., Kaziro, Y., and Itoh, H. (1997) C-terminal mutation of G protein beta subunit affects differentially extracellular signal-regulated kinase and c-Jun N-terminal kinase pathways in human embryonal kidney 293 cells. *J. Biol. Chem.* **272**, 7602–7607.

112. Kumar, C. C. (1998) Signaling by integrin receptors. *Oncogene* **17**, 1365–1373.

113. Giancotti, F. G. and Ruoslahti, E. (1999) Integrin signaling. *Science* **285**, 1028–1032.

114. Li, J., Lin, M. L., Wiepz, G. J., Guadarrama, A. G., and Bertics, P. J. (1999) Integrin-mediated migration of murine B82L fibroblasts is dependent on the expression of an intact epidermal growth factor receptor. *J. Biol. Chem.* **274**, 11,209–11,219.

115. Moro, L., Dolce, L., Cabodi, S., et al. (2002) Integrin-induced epidermal growth factor (EGF) receptor activation requires c-Src and p130Cas and leads to phosphorylation of specific EGF receptor tyrosines. *J. Biol. Chem.* **277**, 9405–9414.

116. Short, S. M., Talbott, G. A., and Juliano, R. L. (1998) Integrin-mediated signaling events in human endothelial cells. *Mol. Biol. Cell.* **9**, 1969–1980.

117. Cazaubon, S., Chaverot, N., Romero, I. A., et al. (1997) Growth factor activity of endothelin-1 in primary astrocytes mediated by adhesion-dependent and -independent pathways. *J. Neurosci.* **17**, 6203–6212.

118. Schlaepfer, D. D. and Hunter, T. (1996) Evidence for in vivo phosphorylation of the Grb2 SH2-domain binding site on focal adhesion kinase by Src-family protein-tyrosine kinases. *Mol. Cell. Biol.* **16**, 5623–5633.

119. Moro, L., Venturino, M., Bozzo, C., et al. (1998) Integrins induce activation of EGF receptor: role in MAP kinase induction and adhesion-dependent cell survival. *EMBO J.* **17**, 6622–6632.

120. Cohen, S. (1960) Purification of a nerve-growth promoting protein from the mouse salivary gland and its neuro-cytotoxic antiserum. *Proc. Natl. Acad. Sci. USA* **46**, 302–311.

121. Fisher, D. A. and Lakshmanan, J. (1990) Metabolism and effects of epidermal growth factor and related growth factors in mammals. *Endocr. Rev.* **11**, 418–42.

122. Murdoch, G. H., Potter, E., Nicolaisen, A. K., Evans, R. M., and Rosenfeld, M. G. (1982) Epidermal growth factor rapidly stimulates prolactin gene transcription. *Nature* **300**, 192–194.

123. Benveniste, R., Speeg, K. V., Jr., Carpenter, G., Cohen, S., Lindner, J., and Rabinowitz, D. (1978) Epidermal growth factor stimulates secretion of human chorionic gonadotropin by cultured human choriocarcinoma cells. *J. Clin. Endocrinol. Metab.* **46**, 169–172.

124. Morrish, D. W., Bhardwaj, D., Dabbagh, L. K., Marusyk, H., and Siy, O. (1987) Epidermal growth factor induces differentiation and secretion of human chorionic gonadotropin and placental lactogen in normal human placenta. *J. Clin. Endocrinol. Metab.* **65,** 1282–1290.

125. Rashed, S. M. and Patel, T. B. (1991) Regulation of hepatic energy metabolism by epidermal growth factor. *Eur. J. Biochem.* **197,** 805–813.

126. Rall, L. B., Scott, J., Bell, G. I., et al. (1985) Mouse prepro-epidermal growth factor synthesis by the kidney and other tissues. *Nature* **313,** 228–231.

127. Holbro, T. and Hynes, N. E. (2004) ErbB receptors: directing key signaling networks throughout life. *Ann. Rev. Pharmacol. Toxicol.* **44,** 195–217.

128. Yarden, Y. (2001) The EGFR family and its ligands in human cancer. Signalling mechanisms and therapeutic opportunities. *Eur. J. Cancer* **37(Suppl 4),** S3–8.

129. Baulida, J., Kraus, M. H., Alimandi, M., Di Fiore, P. P., and Carpenter, G. (1996) All ErbB receptors other than the epidermal growth factor receptor are endocytosis impaired. *J. Biol. Chem.* **271,** 5251–5257.

130. Miettinen, P. J., Berger, J. E., Meneses, J., et al. (1995) Epithelial immaturity and multiorgan failure in mice lacking epidermal growth factor receptor. *Nature* **376,** 337–341.

131. Sibilia, M., Steinbach, J. P., Stingl, L., Aguzzi, A., and Wagner, E. F. (1998) A strain-independent postnatal neurodegeneration in mice lacking the EGF receptor. *EMBO J.* **17,** 719–731.

132. Threadgill, D. W., Dlugosz, A. A., Hansen, L. A., et al. (1995) Targeted disruption of mouse EGF receptor: effect of genetic background on mutant phenotype. *Science* **269,** 230–234.

133. Wong, R. W. (2003) Transgenic and knock-out mice for deciphering the roles of EGFR ligands. *Cell. Molec. Life Sci.* **60,** 113–118.

134. Sealy, L., Privalsky, M. L., Moscovici, G., Moscovici, C., and Bishop, J. M. (1983) Site-specific mutagenesis of avian erythroblastosis virus—Erb-B is required for oncogenicity. *Virology* **130,** 155–178.

135. Yamamoto, T., Hihara, H., Nishida, T., Kawai, S., and Toyoshima, K. (1983) A new avian erythroblastosis virus, Aev-H, carries Erbb gene responsible for the induction of both erythroblastosis and sarcomas. *Cell* **34,** 225–232.

136. Rowinsky, E. K. (2004) The erbB family: Targets for therapeutic development against cancer and therapeutic strategies using monoclonal antibodies and tyrosine kinase inhibitors. *Ann. Rev. Med.* **55,** 433–457.

137. Slamon, D. J., Godolphin, W., Jones, L. A., et al. (1989) Studies of the Her-2/Neu proto-oncogene in human-breast and ovarian-cancer. *Science* **244,** 707–712.

138. Klapper, L. N., Glathe, S., Vaisman, N., Hynes, N. E., Andrews, G. C., Sela, M., and Yarden, Y. (1999) The ErbB-2/HER2 oncoprotein of human carcinomas may function solely as a shared coreceptor for multiple stroma-derived growth factors. *Proc. Natl. Acad. Sci. USA* **96,** 4995–5000.

139. Difiore, P. P., Pierce, J. H., Kraus, M. H., Segatto, O., King, C. R., and Aaronson, S. A. (1987) Erbb-2 is a potent oncogene when overexpressed in Nih/3t3 cells. *Science* **237,** 178–182.

140. Plowman, G. D., Culouscou, J. M., Whitney, G. S., et al. (1993) Ligand-specific activation of Her4/P180(Erbb4), a fourth member of the epidermal growth-factor receptor family. *Proc. Natl. Acad. Sci. USA* **90,** 1746–1750.

141. Kraus, M. H., Issing, W., Miki, T., Popescu, N. C., and Aaronson, S. A. (1989) Isolation and characterization of Erbb3, a third member of the Erbb/epidermal growth-factor receptor family—evidence for overexpression in a subset of human mammary-tumors. *Proc. Natl. Acad. Sci. USA* **86,** 9193–9197.

142. Holgate, S. T., Lackie, P. M., Davies, D. E., Roche, W. R., and Walls, A. F. (1999) The bronchial epithelium as a key regulator of airway inflammation and remodelling in asthma. *Clin. Exp. Allergy* **29,** 90–95.

143. Qian, Q., Harris, P. C., and Torres, V. E. (2001) Treatment prospects for autosomal-dominant polycystic kidney disease. *Kidney Int.* **59,** 2005–2022.

144. Shu, H. K. G., Pelley, R. J., and Kung, H. J. (1991) Dissecting the activating mutations in V-Erbb of avian erythroblastosis virus strain-R. *J. Virol.* **65,** 6173–6180.

145. Paez, J. G., Janne, P. A., Lee, J. C., et al. (2004) EGFR mutations in lung cancer: correlation with clinical response to gefitinib therapy. *Science* **304,** 1497–1500.

146. Pao, W., Miller, V., Zakowski, M., et al. (2004) EGF receptor gene mutations are common in lung cancers from "never smokers" and are associated with sensitivity of tumors to gefitinib and erlotinib. *Proc. Natl. Acad. Sci. USA* **101,** 13,306–13,311.

147. Sordella, R., Bell, D. W., Haber, D. A., and Settleman, J. (2004) Gefitinib-sensitizing EGFR mutations in lung cancer activate anti-apoptotic pathways. *Science* **305,** 1163–1167.

148. Lee, J. W., Soung, Y. H., Kim, S. Y., et al. (2005) Absence of EGFR mutation in the kinase domain in common human cancers besides non-small cell lung cancer. *Int. J. Cancer* **113,** 510–511.

# 2

## Purification and Assay of Kinase-Active EGF Receptor From Mammalian Cells by Immunoaffinity Chromatography

Gregory J. Wiepz, Arturo G. Guadaramma, David L. Fulgham, and Paul J. Bertics

### Summary

The epidermal growth factor (EGF) receptor possesses intrinsic protein–tyrosine kinase activity, and both overexpressed wild-type and mutated forms have been associated with many types of cancers. Therefore, understanding the mechanisms that modulate receptor activity and function is essential to the development of treatments for many of these cancers. However, to address this issue by either conventional or high-throughput screening methods requires the availability of large amounts of highly purified and active EGF receptor.

The technique described in this chapter utilizes immunoaffinity chromatography, which allows for the isolation of highly purified and active preparations of EGF receptor. By immobilizing an antibody that recognizes the ligand-binding domain of the receptor to Sepharose beads, the receptor can be eluted specifically from the antibody by the addition of EGF. This association establishes a unique interaction that ensures the isolation of a highly enriched preparation of EGF receptor. This protocol allows for the purification of large or small batches of receptor that retain their kinase activity. Additionally, this chapter reports on the subsequent steps necessary to characterize the receptor: kinase activity, mass, purity, and the ability of the receptor to undergo autophosphorylation.

**Key Words:** Epidermal growth factor (EGF) receptor; tyrosine kinase; autophosphorylation; immunoaffinity chromatography; receptor purification.

## 1. Introduction

The ability to study enzymes that have been isolated from their natural cellular environment has permitted the extensive evaluation and characterization of the intra- and intermolecular mechanisms regulating enzymatic activity. For example, understanding the internal regulatory mechanisms of the human EGF

From: *Methods in Molecular Biology, vol. 327: Epidermal Growth Factor: Methods and Protocols*
Edited by: T. B. Patel and P. J. Bertics © Humana Press Inc., Totowa, NJ

receptor expressed by many types of cells has led to the generation of numerous potent and highly specific therapeutics for treating various human cancers *(1)*. (In this regard, the reader is referred to Chapter 14 regarding the clinical significance of developing agents that target the EGF receptor.)

The EGF receptor is a large transmembrane glycoprotein that resides in the plasma membrane of a wide range of mammalian cells and possesses intrinsic tyrosine kinase activity *(2)*. Upon activation by several different ligands, e.g., EGF, transforming growth factor-α, or amphiregulin, the receptor undergoes autophosphorylation on numerous tyrosine residues and tyrosine phosphorylates multiple substrates *(3)*. The phosphorylated tyrosine residues in the receptor serve to regulate receptor kinase activity and trafficking and can act as potential docking sites for many cytoplasmic signaling proteins such as Grb2 and phospholipase C (PLC)γ *(4–6)*.

The ability to investigate substrate, inhibitor, and effector protein interactions with purified receptor in vitro has allowed for a more detailed analysis of receptor function. In this regard, multiple in-frame deletions of the EGF receptor have been recently identified in non-small-cell lung cancer patients who were nonsmokers *(7)*. These mutations lead to an increase in receptor phosphorylation on tyrosines 992 and 1068, which promotes the increased activation of downstream mediators, such as AKT and STAT5 *(8)*. This series of mutations also renders the receptor and the tumor susceptible to a class of tyrosine kinase inhibitors that target the ATP-binding pocket of the EGF receptor (i.e., gefitinib, iressa) *(8,9)*. To understand how these mutations alter the ATP-binding site, thereby rendering them susceptible to specific drugs, requires an evaluation of the interaction of the drug with purified receptor.

The method that we employ to isolate functional EGF receptor involves the use of immunoaffinity chromatography *(10)*. Once the receptor is solubilized from the cell membrane, it is bound by a Sepharose-linked mouse monoclonal antibody (anti-EGF receptor—clone 528) that recognizes an epitope in the ligand-binding domain. This specific interaction allows for the elution of the receptor from the antibody using recombinant human EGF (rhEGF), i.e., the rhEGF competes with the immobilized antibody for binding to the receptor. The receptor, which remains functional and bound to EGF, is displaced from the resin, and upon addition of the appropriate components (metal ions, ATP, exogenous peptides), its tyrosine kinase activity can be readily measured *(11)*.

Following elution of the receptor, the preparation is stored at –80°C in the presence of several stabilizing components. The receptor is very stable at this temperature and remains active for at least 1 yr. Following purification, the receptor is quantified regarding mass, purity, kinase activity, and the ability to autophosphorylate. Although this method is described for large-scale purification of EGF receptor, it works equally well on a smaller scale to isolate multiple mutants/variants of the receptor for characterization.

The ability to purify the EGF receptor is critical to the development of drugs that specifically target the EGF receptor kinase and/or protein interaction motifs. Additionally, to better identify and characterize inhibitors directed toward the kinase activity of the EGF receptor, methods that employ high-throughput screening can be facilitated by the ready availability of purified receptor *(12)*. (The reader is directed to Chapter 3 by Lafky et al. for a discussion of other receptor-quantification methods, such as those required for the detection of soluble EGF receptor forms.)

## 2. Materials

### 2.1. Receptor Purification

1. Mammalian cells expressing EGF receptors (e.g., A431 cells).
2. Monoclonal antibody 528 which specifically recognizes the ligand binding site of the human EGF receptor: LabVision, CA (MS-268).
3. Recombinant human EGF: Upstate Biologicals Inc. (01-407).
4. CnBr-activated sepharose 4B beads—Sigma (C-9142).
5. Receptor buffer (20 m$M$ hydroxyethyl piperazine ethane sulfonate [HEPES], pH 7.4, 130 m$M$ NaCl, 1 m$M$ ethylenediaminetetraacetic acid [EDTA], 1 m$M$ dithiothreitol [DTT], 10% glycerol, 0.05% Triton X-100).
6. Homogenization buffer: 40 m$M$ HEPES, pH 7.4, 10 m$M$ ethylene glycol-bis(2 aminoethylether)-$N,N,N',N'$-tetraacetic acid [EGTA], 2% Triton X-100, 20% glycerol.
7. Cell harvesting buffer (10 m$M$ Na$_2$HPO$_4$, pH 7.4, 145 m$M$ NaCl, 2.5 m$M$ EGTA, 2.5 m$M$ EDTA).
8. Receptor buffer/1 $M$ NaCl: receptor buffer, 1 $M$ NaCl.
9. Receptor buffer/1 $M$ urea: receptor buffer, 1 $M$ urea.
10. Receptor buffer/8 $M$ urea: receptor buffer, 8 $M$ urea.
11. Elution buffer: receptor buffer, 25 µg/mL rhEGF.
12. Bead storage buffer: receptor buffer, 0.05% sodium azide.
13. Coupling buffer: 100 m$M$ NaHCO$_3$, 50 m$M$ NaCl.
14. Blocking buffer: 0.5 $M$ ethanolamine in ddH$_2$O.
15. Protease/phosphatase buffer: 2 m$M$ Na$_3$VO$_4$, 100 m$M$ NaF, 10 m$M$ p-nitrophenyl phosphate, 1% aprotinin, 2 m$M$ DTT, 0.2 m$M$ leupeptin, 8 m$M$ benzamidine (*see* **Note 1**).

### 2.2. Assay Buffers

1. Standard assay mix: 20 m$M$ HEPES, pH 7.4, 0.2 m$M$ Na$_3$VO$_4$, 20 m$M$ MgCl$_2$, 8 m$M$ MnCl$_2$, 2% (v/v) aprotinin, 0.1 m$M$ leupeptin.
2. Angiotensin II mix: 20 m$M$ HEPES, pH. 7.4, 3 m$M$ angiotensin II, 50% standard assay mix.
3. ATP mix: 20 m$M$ HEPES, pH 7.4, 30 µ$M$ ATP, 25% standard assay mix, 0.75 µCi [$\gamma^{32}$P]ATP/µL.
4. 20% trichloroacetic acid (TCA) in ddH$_2$O.

5. Autophosphorylation assay mix: 20 m$M$ HEPES, pH 7.4, 2 μ$M$ ATP, 10 m$M$ MgCl$_2$, 4 m$M$ MnCl$_2$, 100 μ$M$ Na$_3$VO$_4$, 1% (v/v) aprotinin, 0.1 m$M$ leupeptin, 0.3 μCi [γ$^{32}$P]ATP/μL.
6. Sample buffer: 20 m$M$ Tris pH 8.0, 1.5 m$M$ EDTA, 20 m$M$ DTT, 2% sodium dodecyl sulfate [SDS], 20% glycerol, 0.2% bromophenol blue.
7. Sample buffer without DTT: same recipe as above without the DTT and without EDTA.

### 2.3. Additional Key Reagents

1. ATP: Adenosine 5'-triphosphate, magnesium salt (Sigma, A9187).
2. Angiotensin II: human synthetic, acetate salt (Sigma, A9525).
3. [γ$^{32}$P]ATP: Amersham Pharmacia (PB10168).
4. Phosphocellulose paper (P81): Whatman (3698-915).
5. Polyacrylamide gel electrophoresis (PAGE) staining reagents: Sypro Orange, Molecular Probes (S6650).
6. All chemicals unless otherwise stated are purchased from Sigma.

### 2.4. Equipment

1. Refrigerated centrifuge: table top (3200$g$; e.g., Beckman Allegra 6R).
2. Centrifuge: super speed (capable of 20,000$g$; e.g., Sorval RC5B).
3. Centrifuge: microfuge (Eppendorf 5415 or 5420).
4. Centrifuge: radioactive microfuge (Eppendorf 5415).
5. Sonicator: Branson Sonifier-250 with microtip.
6. X-ray film developer or a phosphoimager.
7. Polyacrylamide gel electrophoresis (PAGE) equipment.
8. Gel imaging and quantifying equipment.
9. Scintillation counter and scintillation fluid.

## 3. Methods

### 3.1. Cell Culture

Depending on the amount of receptor required, the cell type and cell culture method will vary. To isolate a large amount of wild-type receptor, we prepare approx 5 g of an A431 cell pellet. To achieve this yield, we culture about 250, 15-cm tissue culture plates in groups of 50 plates over a period of several months.

### 3.1.1. Cell Splitting and Culturing

As the cells approach confluence, they are harvested by rinsing the plate with the cell harvesting buffer and then incubating the plates in the same buffer for 10–15 min at 4°C. The cells are then harvested by scraping with a hard edge scraper. The cells are collected, combined into 50-mL polypropylene centrifuge tubes, and centrifuged at 500$g$ (~1500 rpm in a tabletop refrigerated centrifuge) for 10 min at 4°C. The supernatant is removed, and the pellets are frozen at –80°C (*see* **Note 2**).

## 3.2. Bead Preparation

This procedure requires 24 h of preparation time. All washes of the beads will occur in a 15-mL conical polypropylene tube centrifuged at 500*g* (1500 rpm in a tabletop refrigerated centrifuge) for 5 min at 4°C. This protocol is specifically designed to prepare 1 g of CNBr-activated Sepharose beads that will swell to a slurry volume of 3.5 mL, which in turn will then be coupled with 10 mg of anti-EGF receptor antibody. Smaller volumes can be scaled down to accommodate fewer beads and less antibody (*see* **Note 3**).

1. Lyophylized CNBr-activated beads are suspended in 40 mL of 0.1 *M* HCl for 1 h to swell and remove the stabilizer. This procedure is performed in a 50-mL graduated cylinder. Invert the beads five times; after settling, the buffer is poured off, which contains the fines, fragments of beads, and debris. Resuspend the beads in 40 mL of buffer again, and after they settle pour off the buffer. Repeat this step three times. Resuspend the beads in coupling buffer (10 mL) and transfer to a 15-mL conical centrifuge tube.
2. Wash the beads two times by centrifugation (500*g*, 5 min) with 10 mL of coupling buffer.
3. After the final wash add an equal volume of coupling buffer to the bead volume (~3.5 mL). Add the antibody solution (approx 10 mg) directly to the beads. Incubate at room temperature for 2 h with constant inversion.
4. Centrifuge the beads (500*g*, 5 min) and retain the supernatant (to check for antibody coupling efficiency).
5. Wash the beads three times with 10 mL of the coupling buffer and discard the supernatants.
6. Block the unbound active sites on the beads with three bead volumes of blocking buffer (0.5 *M* ethanolamine) overnight with constant inversion at 4°C.
7. The next day and before each use, wash the beads 10 times with 10 mL of receptor buffer.

After each purification, the beads are washed with 10 mL of receptor buffer/ 8 *M* urea, followed by 10 washes of receptor buffer (10 mL), and stored in bead storage buffer (10 mL) at 4°C for up to 1 yr.

## 3.3. Receptor Isolation and Elution

To isolate the receptor from the plasma membrane, the cells are sonicated in the presence of Triton X-100, protease, and phosphatase inhibitors. The receptor is then combined with the beads so that it may bind to the agaose bound antibody. Following the incubation for the specific interaction between the receptor and the antibody, the beads are washed under stringent conditions (1 *M*

NaCl, 1 *M* urea) to remove all nonspecifically bound cell products. Finally, the receptor is displaced from the antibody with EGF, which interacts with the same binding site as the antibody.

1. Upon thawing, the cells are combined with an equal volume of homogenization buffer that includes the protease inhibitor buffer (*see* **Note 4**).
2. The cell solution is then disrupted using a probe sonicator for 3 × 20 s bursts at a setting of 3 (30–50% output; *see* **Note 5**).
3. Centrifuge the cell homogenate at 20,000*g* for 15 min at 4°C. Retain the supernatant, which is where the receptor will predominantly reside. Discard the pellet.
4. Add the supernatant to the prepared beads and incubate overnight at 4°C with constant inversion (*see* **Note 6**).
5. After the overnight incubation, the beads are washed according to the following protocol. All centrifugations are 500*g* for 5 min performed in a 15-mL centrifuge tube. The volume of each of the washes is 15 mL minus the volume of the beads (e.g., fill the tube up for each wash).
6. Centrifuge the beads and collect and retain the supernatant, which will contain any unbound receptor (*see* **Note 7**).
7. Wash the beads 10 times with receptor buffer.
8. Wash the beads three times with receptor buffer /1 *M* NaCl.
9. Wash the beads three times with receptor buffer.
10. Wash the beads three times with receptor buffer/1 *M* NaCl.
11. Wash the beads three times with receptor buffer.
12. Wash the beads three times with receptor buffer /1 *M* urea.
13. Wash the beads five times with receptor buffer.

### 3.3.2. Receptor Elution

To elute the receptor from the beads, add an equal volume of the elution buffer and mix by constant inversion at room temperature for 25 min. Centrifuge the beads and keep the supernatant, which will contain the receptor. Repeat this step and combine the two elution supernatants (*see* **Note 8**).

### 3.3.3. Receptor Storage

Freeze the supernatants in small aliquots (20–500 µL) and store at –80°C (*see* **Note 9**).

### 3.3.4. Bead Storage

The beads are then washed with receptor buffer/8 *M* urea (*see* **Note 10**). Next, the beads are washed 10 times with receptor buffer and finally suspended in twice the bead volume of bead storage buffer. The beads can be used for up to 1 yr after preparation or for a total of six separate purifications. However, this will depend on the quality of the bead preparation, storage, and usage. Additionally, the receptor yield should be evaluated after every purification, as described below (*see* **Note 11**).

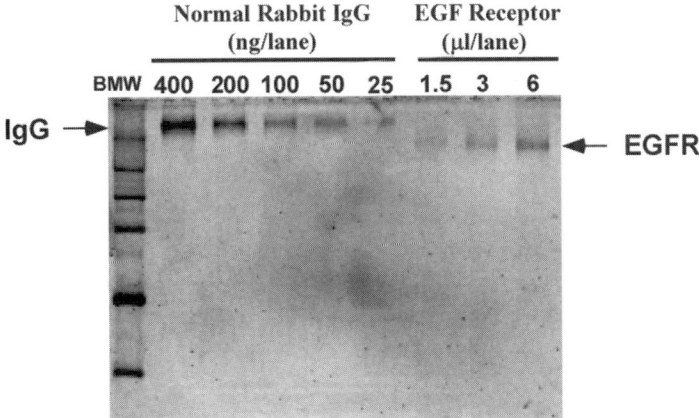

Fig. 1. Determination of epidermal growth factor (EGF) receptor mass and purity. Various amounts of the purified EGF receptor and rabbit IgG were separated on a 10% sodium dodecyl sulfate-polyacrylamide gel electrophoresis (SDS-PAGE). The gel was processed as described and stained with Sypro orange, and the bands were visualized on a Hitachi FMBIO II laser scanner and quantified using the National Institutes of Health Software Image. The data were then processed in Microsoft Excel to produce a standard curve for the IgG to correlate with the EGF receptor values.

## 3.4. Receptor Characterization

Quantification of the purified receptor involves determining three separate parameters: receptor mass and purity, kinase activity, and the ability to undergo autophosphorylation. From these data the quality and quantity of the receptor can be established.

### 3.4.1. Measurement of EGF Receptor Mass

Because the receptor was eluted with buffer containing 25 µg/mL of rhEGF, the mass must be estimated by gel electrophoresis comparing the receptor mass to a protein standard, preferably of similar molecular weight. For this determination, a standard solution of rabbit IgG is used. Because the IgG is made up of heavy and light chains, when prepared in the absence of a reducing agent the molecule will remain intact and migrate at a similar rate to EGF receptor in a 10% polyacrylamide gel (*see* **Fig. 1**). This procedure also gives you an estimate of purity of the receptor.

1. Prepare a 10% polyacrylamide gel.
2. Prepare the IgG standards so that you load 400, 200, 100, 50, 25 ng/well. The volume will depend on the type of gel system and comb size that you use. These standards are prepared in sample buffer that does not contain any reducing agents (DTT) or EDTA.

3. Prepare dilutions of purified EGF receptor (e.g., 6, 3, 1.5 µL of receptor/well).
4. Run the gel and stain (*see* **Note 12**).
5. Quantify the bands for the receptor and the IgG standard by densitometry and determine the protein mass for the EGF receptor per microliter (*see* **Note 13**).

### 3.4.2. Measurement of Receptor Kinase Activity

Kinase activity is determined by quantifying the ability of the purified receptor to phosphorylate angiotensin II under very specific assay conditions. Although there are a number of new techniques to assess the activity of the receptor *(12)*, the method that we employ is the transfer of a radioactive phosphate from $\gamma^{32}$P-ATP to angiotensisn II. Although fairly simple to conduct, this assay does require the use of radioactive materials and the associated equipment and regulatory clearance.

This assay is performed by adding various concentrations of the receptor with the appropriate assay components under very specific conditions (i.e., time and temperature). The timing of the assay is extremely critical, as it is important to assess receptor kinase activity using linear initial velocity conditions *(12)*. Modifying the concentration of the receptor, substrate, ATP, or the time and the temperature will alter the measured activity of the receptor.

**Steps 1–3** are performed on ice.

1. Make the appropriate dilutions of the receptor in receptor buffer. The receptor must be diluted so that each tube will receive exactly 20 µL. Maintain the tubes on ice at all times (*see* **Table 1**) (*see* **Note 14**).
2. Aliquot to all tubes (1.5-mL polypropylene centrifuge tubes) 20 µL of angiotensin II.
3. Add the receptor to the appropriate tubes (20 µL) or control solution (receptor buffer)
4. Warm the ATP mixture in a 30°C water bath for 5 min.
5. Each tube must undergo a 2 min prewarming step prior to receiving the warm ATP mixture to initiate the reaction, which is then incubated at 30°C for exactly 5 min. To facilitate the completion of the assay, the tubes are staggered by 15–30 s so that multiple tubes can be incubating at the same time but are all stopped at the specified time. (*see* **Table 1**).
6. The reactions are terminated by the addition of 20 µL of 20% trichloroacetic acid (TCA) and placed at 4°C.
7. Upon completion of the assay, each tube is centrifuged in a microfuge on max (>10,000 rpm) for 5 min at room temperature (*see* **Note 15**).
8. To allow the angiotensin II to bind to the phosphocelluose paper, the TCA must be diluted. The supernatant (50 µL) is combined with ddH$_2$O (50 µL) in a separate tube, and 90 µL of this solution is spotted onto a 1 × 1 in. piece of phosphocellulose paper that has been prelabeled with the tube number in pencil/pen/marker (*see* **Note 16**).

**Table 1**
**Typical Protocol for an Epidermal Growth Factor Receptor (EGFR) Assay**

| Tube | 50% EGFR (µL) | 25% EGFR (µL) | 12.5% EGFR (µL) | Receptor buffer (µL) | 3 mM AII (µL) | 30 µM ATP (µL) | 20% TCA (µL) | Start (min) | Stop (min) |
|---|---|---|---|---|---|---|---|---|---|
| 1 | 20 | — | — | — | 20 | 20 | 20 | 2 | 7 |
| 2 | 20 | — | — | — | 20 | 20 | 20 | 2.25 | 7.25 |
| 3 | — | 20 | — | — | 20 | 20 | 20 | 2.5 | 7.5 |
| 4 | — | 20 | — | — | 20 | 20 | 20 | 2.75 | 7.75 |
| 5 | — | — | 20 | — | 20 | 20 | 20 | 3 | 8 |
| 6 | — | — | 20 | — | 20 | 20 | 20 | 3.25 | 8.25 |
| 7 | — | — | — | 20 | 20 | 20 | 20 | 3.5 | 8.5 |
| 8 | — | — | — | 20 | 20 | 20 | 20 | 3.75 | 8.75 |
| Total | 40 | 40 | 40 | 40 | 160 | 160 | 160 | | |

9. The paper is allowed to dry and is then washed in a large volume (~300 µL) of cold 0.5% phosphoric acid three times for 10 min with constant, gentle rocking.

10. The filter papers are air dried, added to 5 mL of scintillation fluid, and the activity on each paper is determined by scintillation counting (*see* **Note 17**).

### 3.4.2.1. SAMPLE CALCULATIONS (**TABLE 2**)

#### 3.4.2.1.1. Determination of ATP-Specific Activity

1. ATP concentration—the total concentration of ATP present in the ATP solution.
2. Total counts—the number of dpm per volume of the ATP solution (i.e., 5 µL) (*see* **Note 18**).
3. Specific activity—the number of dpm/pmol of ATP. This number is used to convert the number of dpm into the amount of phosphate incorporated into the substrate (angiotensin II).

#### 3.4.2.1.2. Determination of Kinase Activity

1. Disintegrations per minute (DPM)—actual uncorrected counts.
2. DPM corrected—DPM minus the counts present in the blank tubes (average background).
3. pmol $^{32}$P—DPM corrected divided by the specific activity. This value is the amount of phosphate that was transferred to the substrate (angiotensin II).
4. pmol/assay—pmol$^{32}$P incorporated multiplied by the % reaction counted.
5. % reaction counted—because only a fraction of the actual assay reaction volume is counted, the cpm needs to be corrected. From the original assay volume, 50 µL is removed and combined with 50 µL of water to dilute the concentration of TCA, which permits the binding of the substrate to the P-81 paper. From this diluted sample, 90 µL is applied to the paper. Therefore, the amount of sample counted is 56% of the original tube ($0.9 \times 0.625 = 0.56$).
6. Velocity (pmol/min)—pmol/assay divided by the time (min), usually 5 min.
7. Average velocity—average of the duplicates for each dilution.
8. Velocity pmol/min/µL—average velocity divided by the number of µL of receptor originally used.

### 3.4.3. Receptor Autophosphorylation (**Fig. 2**)

Because the EGF receptor is known to autophosphorylate in vivo, evaluation of this parameter reflects the degree to which the receptor has remained intact following purification. This assay is similar to the standard kinase assay, with the exclusion of any additional substrate. Following the reaction, varying amounts of the receptor are separated by SDS-PAGE and exposed to radiographic film for the appropriate amount of time (*see* **Note 19**).

1. Add 10 µL of the purified receptor at 4°C with 10 µL of the Autophosphorylation buffer.
2. Incubate on ice for 10 min and stop the reaction with 20 µL of sample buffer.
3. Prepare a 10% PAGE and load various volumes of the reaction (e.g., 20, 10, 5, 2.5 µL, which is equivalent to 5, 2.5, 1.25, and 0.63 µL of receptor).

# Table 2
## Representative Data from Epidermal Growth Factor Receptor Kinase Assay[a]

Specific activity

ATP concentration = 3.00E-05 M  %Reaction counted = 0.56
μL counted = 0.5          Reaction time (min) = 5

| Total ct | Average ct | dpm/μL | ATP (pmol/μL) | dpm/pmol |
|---|---|---|---|---|
| 904091 | 887661 | 1775322 | 30 | 59177.40 |
| 871231 | | | | |

| Tube | Receptor (μL/tube) | DPM | DPM corrected | pmol $^{32}$P | pmol/assay | Velocity pmol/min | Average velocity | SD | Velocity pmol/min/μL receptor |
|---|---|---|---|---|---|---|---|---|---|
| 1 | 10 | 399383 | 376282 | 6.3585 | 11.3545 | 2.2709 | 2.3073 | 0.0515 | 0.2307 |
| 2 | 10 | 411449 | 388348 | 6.5624 | 11.7186 | 2.3437 | | | |
| 3 | 5 | 191287 | 168186 | 2.8421 | 5.0751 | 1.0150 | 0.9592 | 0.0789 | 0.1918 |
| 4 | 5 | 172787 | 149686 | 2.5294 | 4.5169 | 0.9034 | | | |
| 5 | 2.5 | 94680 | 71579 | 1.2096 | 2.1599 | 0.4320 | 0.5893 | 0.2224 | 0.2357 |
| 6 | 2.5 | 146797 | 123696 | 2.0903 | 3.7326 | 0.7465 | | | |

Background 1 = 24589          Average velocity/μL = 0.2
Background 2 = 21613
Average background = 23101

[a]This analysis was performed in an Excel worksheet.
DPM, disintegrations per minute; SD, standard deviation.

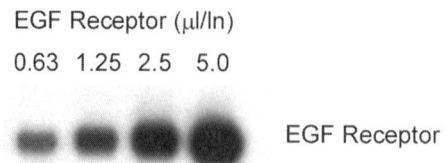

Fig. 2. Epidermal growth factor (EGF) receptor autophosphorylation. The reaction for autophosphorylation was performed as described using purified EGF receptor (43 ng/μL). The radiographic film was exposed for 60 min.

4. Run gel for the appropriate time (*see* **Note 20**).
5. Fix gel with 70% methanol/10% glacial acetic acid for 15 min.
6. Rehydrate gel with ddH$_2$O for 20 min.
7. Wrap gel in plastic wrap and expose to film for 1–2 h or overnight. Alternatively, a phosphoimager can be used.

# 4. Notes

1. As an alternative, protease inhibitor cocktail can be used in place of aprotinin, leupeptin, and benzamidine (i.e., Sigma—Mammalian Protease Inhibitor Cocktail, P8340).
2. These cell pellets are very stable at –80°C and can be stored for more than a year.
3. The hybridoma clone for this antibody can be purchased from American Type Culture Collection (www.atcc.org; HB-8509) and the antibody produced/purified using a hybridoma facility.
4. Make the protease/phosphatase inhibitor buffer just prior to use. Alternatively, a protease inhibitor cocktail can be used (Sigma P8340) with the addition of the phosphatase inhibitors.
5. Always sonicate on ice and allow 1 min between sonications. The probe intensity varies between machines, and the required amount of sonication will need to be determined empirically by checking for intact cells by microscopy.
6. Adjust the volume to 12 mL in a 15-mL tube, or use a smaller tube to decrease the number of beads that stick to the tube.
7. For large preps a considerable amount of receptor will remain in the solution and can be saved and bound to the antibody beads again after the first elution.
8. If there is only a small volume of beads, decrease the tube size to better match the volume.
9. The receptor is commonly stored in multiple small aliquots (25 μL) to avoid repeated freeze/thaws.
10. After centrifugation, the supernatant containing the denatured protein is retained and can be used as a 170 kDa marker for SDS-PAGE.
11. Due to the nature of the procedure, the volume of beads will go down after each use. However, the remaining beads will retain most of their activity through multiple purifications.

12. Several staining procedures are available to quantify the amount of protein present on the gel. We routinely use Sypro Orange (Molecular Probes, Inc.) and visualize the gel in a Laser scanner (Hitachi FMBIO II). Alternatively, depending on the concentration of protein present, other methods can be employed (i.e., Coomassie, silver stain)
13. Various imaging software programs are available to record and quantify the intensity of each band, such as Image (Mac) and Image J (PC), both of which are available from http://rsb.info.nih.gov/nih-image/.
14. Receptor dilution is expressed as a percentage to address a possible range in activity.
15. This procedure will make the centrifuge very radioactive, and appropriate handling procedures must be adhered too.
16. To help spread the 90 µL of assay mixture on the filter paper, the papers are suspended on straight pins that are inserted into a Styrofoam block.
17. To facilitate the drying of the phosphocellulose paper, the individual papers are dried under a heat lamp.
18. The number of dpm is equal to the number of cpm in the scintillation counter that we use with the appropriate scintillation cocktail.
19. Alternatively, this detection can be performed on a phosphoimager.
20. Try not to run the dye front off the gel where a major amount of the radioactivity is located. Stop the gel before the dye front comes off the bottom, cut it off, and discard it appropriately.

## Acknowledgments

This work was supported by grants from the National Institutes of Health (CA105730 and GM53271).

## References

1. Lynch, T. J., Bell, D. W., Sordella, R., et al. (2004) Activating mutations in the epidermal growth factor receptor underlying responsiveness of non-small-cell lung cancer to gefitinib. *N. Engl. J. Med.* **350,** 2129–2139.
2. Carpenter, G. and Cohen, S. (1990) Epidermal growth factor. *J. Biol. Chem.* **265,** 7709–7712.
3. Bertics, P. J. and Gill, G. N. (1985) Self-phosphorylation enhances the protein-tyrosine kinase activity of the epidermal growth factor receptor. *J. Biol. Chem.* **260,** 14,642–14,647.
4. Buday, L. and Downward, J. (1993) Epidermal growth factor regulates p21ras through the formation of a complex of receptor, Grb2 adapter protein, and Sos nucleotide exchange factor. *Cell* **73,** 611–620.
5. Wang, X. J., Liao, H. J., Chattopadhyay, A., and Carpenter, G. (2001) EGF-dependent translocation of green fluorescent protein-tagged PLC-gamma1 to the plasma membrane and endosomes. *Exp. Cell. Res.* **267,** 28–36.
6. Zhu, G., Decker, S. J., and Saltiel, A. R. (1992) Direct analysis of the binding of Src-homology 2 domains of phospholipase C to the activated epidermal growth factor receptor. *Proc. Natl. Acad. Sci. USA* **89,** 9559–9563.

7. Pao, W., Miller, V., Zakowski, M., et al. (2004) EGF receptor gene mutations are common in lung cancers from "never smokers" and are associated with sensitivity of tumors to gefitinib and erlotinib. *Proc. Natl. Acad. Sci. USA* **101,** 13,306–13,311.
8. Sordella, R., Bell, D. W., Haber, D. A., and Settleman, J. (2004) Gefitinib-sensitizing EGFR mutations in lung cancer activate anti-apoptotic pathways. *Science* **305,** 1163–1167.
9. Paez, J. G., Janne, P. A., Lee, J. C., et al. (2004) EGFR mutations in lung cancer: correlation with clinical response to gefitinib therapy. *Science* **304,** 1497–1500.
10. Weber, W., Bertics, P. J., and Gill, G. N. (1984) Immunoaffinity purification of the epidermal growth factor receptor. Stoichiometry of binding and kinetics of self-phosphorylation. *J. Biol. Chem.* **259,** 14,631–14,636.
11. Gronowski, A. M. and Bertics, P. J. (1993) Evidence for the potentiation of epidermal growth factor receptor tyrosine kinase activity by association with the detergent-insoluble cellular cytoskeleton: analysis of intact and carboxy-terminally truncated receptors. *Endocrinology* **133,** 2838–2846.
12. Beebe, J. A., Wiepz, G. J., Guadarrama, A. G., Bertics, P. J., and Burke, T. J. (2003) A carboxyl-terminal mutation of the epidermal growth factor receptor alters tyrosine kinase activity and substrate specificity as measured by a fluorescence polarization assay. *J. Biol. Chem.* **278,** 26,810–26,816.

# 3

## Soluble Epidermal Growth Factor Receptor Acridinium-Linked Immunosorbent Assay

Jacqueline M. Lafky, Andre T. Baron, and Nita J. Maihle

### Summary

This chapter provides basic information for quantifying soluble epidermal growth factor receptor (sEGFR) isoforms in human sera using an acridinium-linked immunosorbent assay (ALISA) developed by Baron and coworkers (*1*). This ALISA has been shown to be specific for epidermal growth factor receptor (EGFR) and the isoforms of EGFR encoded by alternate transcripts (i.e., sEGFRs); it, therefore, does not detect other EGFR/ErbB receptor family members, i.e., ErbB2, ErbB3, or ErbB4. In addition, this ALISA recognizes EGFR and sEGFR isoforms that contain subdomains I–IV of the extracellular domain, but it does not recognize EGFR isoforms lacking subdomain IV. The ALISA described here also may be useful for determining EGFR and sEGFR concentrations in lysates of cultured cells, conditioned culture medium, or tissue/tumor specimens, as well as in other human body fluids such as serum, plasma, ascites, urine, saliva, and cystic fluids.

**Key Words:** Epidermal growth factor receptor; EGFR; soluble EGFR; sEGFR; acridinium; acridinium-linked immunosorbent assay; ALISA; serum; biomarker; cancer.

## 1. Introduction

The *EGFR* proto-oncogene is being studied in many types of human carcinomas, as well as in certain other tumors such as gliomas. This proto-oncogene encodes the epidermal growth factor receptor (EGFR), which is a 170-kDa glycoprotein comprised of an extracellular ligand-binding domain (ECD), a single-pass transmembrane domain, and an intracellular cytoplasmic domain with tyrosine kinase activity (*2*). This receptor is expressed in many human adult tissues (*3,4*) and plays an important physiological role in cell proliferation, differentiation, and survival. Hence, it is not surprising that gene amplification, protein overexpression, and genetic alterations resulting in aberrant

From: *Methods in Molecular Biology, vol. 327: Epidermal Growth Factor: Methods and Protocols*
Edited by: T. B. Patel and P. J. Bertics © Humana Press Inc., Totowa, NJ

EGFR structure and signal transduction contribute to malignant transformation. Numerous investigators have reported *EGFR* gene amplification and/or EGFR protein overexpression in a variety of human tumor-derived cell lines and tissues, such as bladder, brain, breast, cervical, esophageal and gastric, head and neck, ovarian, and pancreatic cancer cells (reviewed in **ref. 5**). Clinically, *EGFR* amplification and/or EGFR overexpression in tumors has been shown to correlate with tumor size, poor tumor differentiation, tumor invasiveness and a high metastatic rate, and poor patient prognosis (e.g., increased disease recurrence and decreased survival), as well as with drug resistance and responsiveness to therapy (reviewed in **ref. 5**).

Soluble EGFR molecules (sEGFR), which contain only the ECD of the receptor, have also been identified (reviewed in **ref. 6**). Human sEGFR proteins identified to date are generated by alternate mRNA splicing/processing events and include 60- *(7)*, 80- *(8)*, 110- *(9)*, and 115-kDa *(10)* isoforms. Although the function(s) of human sEGFRs is an active area of investigation, previous studies with rat and avian sEGFRs have shown that sEGFR isoforms decrease cell proliferation both in vitro *(11,12)* and in vivo *(13)*. The mechanism underlying decreased cell proliferation may involve competitive binding of sEGFR isoforms to growth factors *(14)* and/or kinase inhibition via the ligand-dependent formation of sEGFR and full-length EGFR heterodimers *(15)*. Given the important role of EGFR in modulating cell proliferation and transformation, we, and others, hypothesize that sEGFR concentrations might be altered during oncogenesis, tumor progression, and metastasis. Furthermore, with the ability to quantify serum sEGFR using noninvasive immunoassay methods involving minimal risk to patients, sEGFR isoforms recently have become a major focus of research aimed at evaluating their clinical utility as cancer biomarkers for (1) identifying individuals at increased cancer risk, (2) detecting preclinical cancer, (3) diagnosing a clinically detectable mass, (4) predicting responsiveness to therapy, (5) monitoring responsiveness to therapy, (6) monitoring disease recurrence, and (7) predicting disease outcome. Toward this end, we have developed and validated an exquisitely sensitive acridinium-linked immunosorbent assay that has utility in quantifying sEGFR levels in human body fluids and tissues *(1)*. The characteristics of this assay and a detailed description of the techniques used in the implementation of this assay are the focus of this chapter.

**Table 1** provides a brief summary of the ALISA protocol. The ALISA immunoassay described here is a chemiluminescent direct sandwich immunosorbent assay that utilizes two EGFR ECD-specific mouse monoclonal antibodies (MAbs): a capture antibody ($IgG_{2b}$ isotype) and an acridinium-conjugated antibody ($IgG_{2a}$ isotype). The capture antibody is immobilized to the surface of microplate wells that have been covalently coated with anti-mouse

**Table 1**
**Acridinium-Linked Immunosorbent Assay Protocol Summary**

| Steps | Incubation time and temperature |
|---|---|
| 1. Add 100 μL coating antibodies to each well | Overnight at 4°C |
| 2. Wash | |
| 3. Add 200 μL blocking buffer to each well | 1 h at room temperature |
| 4. Wash | |
| 5. Add 100 μL capture antibody to each well | 2 h at 37°C |
| 6. Wash | |
| 7. Add 100 μL antigen to each well | 2 h at 37°C |
| 8. Wash | |
| 9. Add 100 μL detection antibody to each well | 1 h at 37°C |
| 10. Wash | |
| 11. Read microplate at 430 nm with a luminometer immediately after addition of 100 μL of each chemiluminescent initiator reagents | Room temperature |

$IgG_{2b}$ isotype-specific antibodies. In this immunoassay, specimens containing analytes of known or unknown concentration are incubated in the coated well to allow analyte binding by the capture antibody. The immobilized analyte is then incubated with the acridinium-conjugated antibody to complete the "sandwich." Sequential addition of chemiluminescent initiator reagents, a suitable base (e.g., sodium hydroxide), and oxidant (e.g., hydrogen peroxide), initiates acridinium decomposition, resulting in visible photon emission at a wavelength of approx 430 nm *(16)*. A photomultiplier tube detector in a luminometer detects emitted photons as relative light units (RLUs). The RLUs are directly related to the concentration of EGFR and/or sEGFR in the analytical standard or the specimen.

## 2. Materials

1. 4-(2-Succinimidyl-oxycarbonylethyl)phenyl-10-acridinium-9-carboxylate trifluoromethyl sulfonate (acridinium $C_2$ NHS ester; ASSAY Designs Inc., Ann Arbor, MI). Keep desiccated and store at –20°C. Bring to room temperature before opening.
2. Dry dimethyl formamide.
3. Labeling buffer: 0.2 $M$ sodium phosphate, pH 8.0. Filter and store at room temperature.
4. Quenching buffer: labeling buffer with 10 mg/mL lysine monohydrochloride. Make fresh prior to use.
5. Acridinium storage buffer without bovine serum albumin (BSA): 150 m$M$ NaCl, 0.2 $M$ sodium phosphate, pH 6.3. Filter and store at 4°C.

6. Acridinium storage buffer with BSA: 150 m$M$ NaCl, 0.2 $M$ sodium phosphate, pH 6.3, 0.2% BSA. Filter and store at 4°C.
7. Fast desalting gel filtration columns: HiPrep™ 26/10 desalting, HiTrap™ desalting, or disposable PD-10 desalting columns (Amersham Biosciences, Piscataway, NJ).
8. Test tubes: 50- and 15-mL conical tubes, and 1.5-mL microfuge tubes.
9. Ultracentrifugation filter units: Centricon YM-30 and Ultrafree-MC GV Centrifugal Filter Units (Millipore, Billerica, MA).
10. Coating antibodies: affinity-purified goat α-mouse IgG$_{2b}$ specific polyclonal antibodies.
11. Capture antibody: anti-EGFR MAb EGFR.1 (IgG$_{2b}$ isotype; Lab Vision Corporation, Fremont, CA) *(17)*.
12. Detection antibody: anti-EGFR MAb 528 (IgG$_{2a}$ isotype; Santa Cruz Biotechnology, Inc., Santa Cruz, CA) *(18)*. This MAb will be conjugated to acridinium.
13. Purified EGFR (positive analytical standard; Sigma-Aldrich, St. Louis, MO).
14. Purified negative analytical standard. We use trpEcentrin as our negative analytical standard, which was purified by Baron and coworkers *(19)*.
15. XENOBIND™ white 96-well microplate (XENOPORE Corp., Hawthorne, NJ).
16. Microplate sealers.
17. Multichannel pipettor, single-channel pipettor, repeat pipettor, and/or robot workstation (optional).
18. 10-mL pipets and 1000-, 200-, and 20-µL pipet tips.
19. Manual or automated microplate washer.
20. Coating buffer: 100 m$M$ carbonate/bicarbonate buffer, pH 9.4. Filter and store at room temperature.
21. ALISA blocking buffer (ALBB): 2% BSA, 1 m$M$ ethylene glycol tetraacetic acid (EGTA), 1 m$M$ ethylenediaminetetraacetic acid (EDTA), 0.01% rabbit serum, 0.01% mouse serum, pH 7.4. Filter and store at 4°C.
22. Tween-20 wash buffer (T20WB): 20 m$M$ Tris-HCl buffer, pH 7.4, 400 m$M$ NaCl, 0.05% Tween-20. Store at room temperature.
23. Chemiluminescent initiator solution 1: 0.441% nitric acid, 0.495% $H_2O_2$. Filter and store at room temperature.
24. Chemiluminescent initiator solution 2: 0.25 $M$ NaOH, 0.1875% cetyltrimethylammonium chloride. Filter and store at room temperature.
25. Microplate luminometer.
26. Graphing software package (optional).

## 3. Methods

### 3.1 Acridinium Labeling of Anti-EGFR MAb 528

1. Dilute the acridinium $C_2$ NHS ester to 1 mg/mL in dry dimethyl formamide and store in 10-µg (10-µL) aliquots at –20°C.
2. Exchange the detection antibody, anti-EGFR MAb 528, from carrier solution into labeling buffer using a fast desalting gel filtration column. The labeling buffer must not contain any amines (e.g., Tris or azide).

3. Collect 0.5- or 1.0-mL column fractions. Measure the absorbance of each fraction at 280 nm and combine the protein-containing fractions.
4. Concentrate the protein-containing column fractions to 0.25–1.0 mg/mL using an ultracentrifugation filter unit. Record the final volume.
5. Add enough acridinium $C_2$ NHS ester solution to yield a 40-fold molar excess of the acridinium $C_2$ NHS ester over MAb 528 protein. Vortex gently to mix solution.
6. Incubate the protein-acridinium $C_2$ NHS ester solution for 15 min in the dark at room temperature.
7. Add 20 μL of quenching buffer per 1 μg of acridinium $C_2$ NHS ester used in the protein-acridinium $C_2$ NHS ester solution and incubate for 5 min in the dark at room temperature.
8. Separate unbound acridinium $C_2$ NHS ester and lysine from the acridinium-labeled detection antibody (AC-MAb 528) by chromatography with a fast desalting gel filtration column, while simultaneously exchanging the solution for acridinium storage buffer without BSA.
9. The labeled protein fractions will occur in the first few fractions, and these should be pooled and concentrated using an ultracentrifugation filter unit. Record the final volume.
10. Add an equal volume of acridinium storage buffer with BSA to obtain a final solution containing 1% BSA.
11. Determine the relative light units (RLUs)/μL of the AC-MAb 528 reagent using a luminometer. Aliquot and store the AC-MAb 528 at –70°C.

### *3.2. Soluble EGFR Acridinium-Linked Immunosorbent Assay*

1. Dilute the coating antibodies (goat anti-mouse $IgG_{2b}$ polyclonal antibodies) to 250 μg/mL in 100 m$M$ carbonate/bicarbonate buffer, pH 9.4. Add 100 μL (25 μg/well) of coating antibody solution to each well of a XENOBIND™ white 96-well microplate (*see* **Note 1**).
2. Cover the microplate with a microplate sealer and incubate at 4°C overnight on a rocker platform (*see* **Note 2**).
3. Wash the microplate three times, 300 μL per well, with T20WB (*see* **Notes 3** and **4**).
4. Add 200 μL ALBB to each well of the microplate.
5. Cover the microplate with a microplate sealer and incubate at room temperature for 1 h on a rocker platform.
6. Wash the microplate three times, 300 μL per well, with T20WB.
7. Dilute the capture antibody (anti-EGFR MAb EGFR.1) to 0.5 μg/mL in ALBB. Add 100 μL (0.05 μg/well) capture antibody solution to each well of the microplate.
8. Cover the microplate with a microplate sealer and incubate at 37°C for 2 h in a shaker incubator.
9. Wash the microplate three times, 300 μL per well, with T20WB.
10. Prepare an eight-tube dilution series for both the positive (EGFR standard) and negative (user's protein standard of choice) control antigens. The sEGFR ALISA has a broad linear range (~1 to 4000 fmol/mL); therefore, the first tube of the EGFR standard dilution series and negative control antigen should be diluted to approx

4000 fmol/mL. To prepare the eight-tube dilution series, prepare 500 μL of each analytical standard in the first tube of the dilution series. Transfer 250 μL to the next tube of the dilution series and mix thoroughly. Continue transferring 250 μL to the next tube of the dilution series sequentially.

11. Dilute serum (or other unknown) specimen 1:10 or 1:20 with ALBB. For a 1:10 dilution, pipet 30 μL of serum into 270 μL of ALBB. For a 1:20 dilution, pipet 15 μL of serum into 285 μL of ALBB.

12. Add 100 μL of each analytical standard and serum (or other unknown) specimen, in duplicate, to the designated wells of the microplate.

13. Cover the microplate with a microplate sealer and incubate at 37°C for 2 h in a forced-air environmental shaker incubator.

14. Wash the microplate three times, 300 μL per well, with T20WB.

15. Dilute the detection antibody (acridinium-labeled anti-EGFR MAb 528) to 4000–6000 counts/μL in ALBB. Add 100 μL of detection antibody solution to each well of the microplate.

16. Cover the microplate with a plate sealer and incubate at 37°C for 1 h in a forced-air environmental shaker incubator.

17. Wash the microplate three times, 300 μL per well, with T20WB.

18. Read each well of the microplate with a microplate luminometer for 3 s immediately after sequentially adding 100 μL of chemiluminescent initiator solution 1 and 100 μL of chemiluminescent initiator solution 2 to the well (*see* **Note 5**).

The methods outlined above describe the current working protocol for the ALISA, which was originally developed by Baron and coworkers *(1)*. This ALISA is composed of a complex assortment of reagents and equipment; therefore, failure to follow this protocol as described may cause altered or incomparable results. In addition to the notes referred to above, the reader is encouraged to read **Notes 6–8** to minimize variability in the data.

### 3.3. Evaluating Results

1. Standard concentrations:

   a. Determine the mean RLU for each analytical standard individually. Determine the mean RLU (zero calibrator) and standard deviation for all negative control standards combined.

   b. Plot the mean RLU (*y*-axis) for each EGFR standard in fmol/mL (*x*-axis) and determine the best fit line using linear regression analysis. Graphing software packages are available to simplify this procedure.

   c. Determine the analytical detection limit or lower limit of detection (LLD), which is defined as 2.5 standard deviations (SD) above the zero calibrator (mean RLUs of negative control analyte), and the biological detection limit (BDL), which is defined as 2 SD above the LLD *(20,21)*.

2. Specimen concentrations:

   a. Determine the mean RLU for each serum (or other unknown) specimen.

    b. Serum specimens yielding RLUs below the BDL or above the linear range of the ALISA's standard curve should be reassayed at a lesser or greater dilution, respectively.

    c. For serum (or other unknown) specimens with RLUs within the linear range of the assay, extrapolate the sEGFR concentration from the EGFR standard curve's best-fit line. For each serum (or other unknown) specimen, multiply this concentration by the specimen's dilution factor to obtain the final concentration in fmol/mL.

## 4. Notes

1. White microplates are used to eliminate "cross-talk" between wells, so that photons emitted from one well do not interfere with photons emitted from an adjacent well. The XENOBIND™ microplates are recommended, because these enhanced microplates allow covalent binding of the anti-mouse IgG$_{2b}$ isotype-specific antibodies to the microplate under the conditions described above for coating the plate. Covalent binding allows for increased binding capacity, and the nonreversible nature of covalent binding eliminates antibody displacement caused by either washing procedures or by competition with serum proteins. It is important to keep a record of the microplate lot, since variable results may occur between different lots of microplates. It is good laboratory practice to use the same plate lot for all the serum specimens within a particular study to decrease both intra- and inter-assay variability.

2. We recommend that microplates be placed on a rocker or shaker during each incubation step to ensure thorough mixing of the reagents. Mixing allows for uniform binding of immunoreactive molecules and maintains a constant temperature of reagents within the well.

3. Incomplete washing between incubations may result in increased nonspecific binding and higher background, which consequently leads to higher biological detection limits and lower assay sensitivity. Improper washing may also result in potential cross-contamination between wells; resulting in erroneous calculations of concentrations. We recommend using an automated microplate washer with agitation capability to eliminate cross-contamination between wells and to ensure consistent washing of the plate, thus reducing background noise.

4. If an automatic plate washer is used, the instrument should be flushed with water at the end of each day. If the washing buffer is not removed, the dispensing and aspirating nozzles may become clogged.

5. The injectors of the luminometer should be flushed with water after every day of use. If the chemiluminescent reagents are not removed, the injectors may become clogged, resulting in incorrect volume delivery.

6. A major source of variability in the measurement of sEGFR concentrations arises from differences in laboratory technique between scientists. Individuals must be accurate and consistent in sample handling and in adhering to protocol directions to minimize pipetting errors and sources in intra- and inter-assay variability. Quality control protocols should be developed to assess assay variability as a result of operator differences.

7. All assay reagents must be prepared with care and standardized. New lots of each reagent must be tested and compared to previous lots of the same reagent to verify lot quality. Once a new lot is validated, smaller working aliquots of the reagent should be prepared and stored at −70°C. Additionally, reagents must be kept free of contamination.

8. Pipettors should be routinely calibrated to assure accuracy of delivery volumes. Moreover, an automatic robotic workstation will facilitate standardized pipetting for optimal reproducibility and reduced variability, as well as increased throughput.

## Acknowledgments

This work was supported by National Institutes of Health (NIH) grant KO7 CA76170 to ATB and by the NIH Office of Women's Health Research, and the National Ovarian Cancer Early Detection Research Network (CA 85113), sponsored by the National Cancer Institute. We are grateful for Dr. Jill Reiter's critical review of this manuscript.

## References

1. Baron, A. T., Lafky, J. M., Connolly, D. C., et al. (1998) A sandwich type acridinium-linked immunosorbent assay (ALISA) detects soluble ErbB1 (sErbB1) in normal human sera. *J. Immunol. Methods* **219,** 23–43.

2. Ullrich, A., Coussens, L., Hayflick, J. S., et al. (1984) Human epidermal growth factor receptor cDNA sequence and aberrant expression of the amplified gene in A431 epidermoid carcinoma cells. *Nature* **309,** 418–425.

3. Gullick, W. J. (1991) Prevalence of aberrant expression of the epidermal growth factor receptor in human cancers. *Br. Med. Bull.* **47,** 87–98.

4. Prigent, S. A. and Lemoine, N. R. (1992) The type 1 (EGFR-related) family of growth factor receptors and their ligands. *Prog. Growth Factor Res.* **4,** 1–24.

5. Salomon, D. S., Brandt, R., Ciardiello, F., and Normanno, N. (1995) Epidermal growth factor-related peptides and their receptors in human malignancies. *Crit. Rev. Oncol. Hematol.* **19,** 183–232.

6. Maihle, N. J., Baron, A. T., Barrette, B. A., et al. (2002) EGF/ErbB receptor family in ovarian cancer. *Cancer Res. Treat.* **107,** 247–258.

7. Reiter, J. L. and Maihle, N. J. (1996) A 1.8 kb alternative transcript from the human epidermal growth factor receptor gene encodes a truncated form of the receptor. *Nucleic Acids Res.* **24,** 4050–4056.

8. Ilekis, J. V., Stark, B. C., and Scoccia, B. (1995) Possible role of variant RNA transcripts in the regulation of epidermal growth factor receptor expression in human placenta. *Mol. Reprod. Develop.* **41,** 149–156.

9. Reiter, J. L., Threadgill, D. W., Eley, G. D., et al. (2001) Comparative genomic sequence analysis and isolation of human and mouse alternative EGFR transcripts encoding truncated receptor isoforms. *Genomics* **71,** 1–20.

10. Xu, Y.-H., Richert, N., Ito, S., Merlino, G. T., and Pastan, I. (1984) Characterization of epidermal growth factor receptor gene expression in malignant and normal human cell lines. *Proc. Natl. Acad. Sci. USA* **81,** 7308–7312.

11. Flickinger, T. W., Maihle, N. J., and Kung, H. J. (1992) An alternatively processed mRNA from the avian c-erbB gene encodes a soluble, truncated form of the receptor that can block ligand-dependent transformation. *Mol. Cell. Biol.* **12,** 883–893.

12. Petch, L. A., Harris, J., Raymond, V. W., Blasband, A., Lee, D. C., and Earp, H. S. (1990) A truncated, secreted form of the epidermal growth factor receptor is encoded by an alternatively spliced transcript in normal rat tissue. *Mol. Cell. Biol.* **10,** 2973–2982.

13. Nieto-Sampedro, M. and Broderick, J. T. (1989) A soluble brain molecule related to epidermal growth factor receptor is a mitogen inhibitor for astrocytes. *J. Neurosci. Res.* **22,** 28–35.

14. Cadena, D. L. and Gill, G. N. (1993) Expression and purification of the epidermal growth factor receptor extracellular domain utilizing a polycistronic expression system. *Protein Expr. Purif.* **4,** 177–186.

15. Basu, A., Raghunath, M., Bishayee, S., and Das, M. (1989) Inhibition of tyrosine kinase activity of the epidermal growth factor (EGF) receptor by a truncated receptor form that binds to EGF: role for interreceptor interaction in kinase regulation. *Mol. Cell. Biol.* **9,** 671–677.

16. Weeks, I., Sturgess, M., Brown, R. C., and Woodhead, J. S. (1986) Immunoassays using acridinium esters. *Methods Enzymol.* **133,** 366–387.

17. Waterfield, M. D., Mayes, E. L., Stroobant, P., et al. (1982) A monoclonal antibody to the human epidermal growth factor receptor. *J. Cell. Biochem.* **20,** 149–161.

18. Sato, J. D., Kawamoto, T., Le, A. D., Mendelsohn, J., Polikoff, J., and Sato, G. H., (1983) Biological effects in vitro of monoclonal antibodies to human epidermal growth factor receptors. *Mol. Biol. Med.* **1,** 511–529.

19. Baron, A. T., Greenwood, T. M., Bazinet, C. W., and Salisbury, J. L., (1992) Centrin is a component of the pericentriolar lattice. *Biol. Cell* **76,** 383–388.

20. Rodbard, D. (1978) Statistical estimation of the minimal detectable concentration ("sensitivity") for radioligand assays. *Anal. Biochem.* **90,** 1–12.

21. Spencer, C. A., Takeuchi, M., Kazarosyan, M., MacKenzie, F., Beckett, G. J., and Wilkinson, E. (1995) Interlaboratory/intermethod differences in functional sensitivity of immunometric assays of thyrotropin (TSH) and impact on reliability of measurement of subnormal concentration of TSH. *Clin. Chem.* **41,** 367–374.

# 4

## Analysis of EGF Receptor Interactions With the α Subunit of the Stimulatory GTP Binding Protein of Adenylyl Cyclase, $G_s$

### Deepti Chaturvedi, Francis Edwin, Hui Sun, and Tarun B. Patel

### Summary

Besides stimulating the mitogen-activated protein kinase, phospholipase Cγ, and phosphatidylinositol 3-kinase cascades, in certain tissues and cells such as the heart, partotid gland, and luteal cells, activation of the epidermal growth factor (EGF) receptor also stimulates second-messenger systems that involve the heterotrimeric G proteins. For instance, in the heart EGF increases contractility and heart rate by elevating cellular cyclic adenosine monophosphate (cAMP) levels. This is the result of EGF-elicited activation of adenylyl cyclase via the stimulatory guanosine 5'-triphosphate (GTP)-binding protein $G_s$. In this context, the single transmembrane EGF receptor acts like a heptahelical G protein-coupled receptor. Here we have described the methods used to study interactions between the EGF receptor and heterotrimeric G proteins. Moreover, we have also described how the stoichiometry of EGF receptor association with the α subunit of $G_s$ can be monitored in vitro. Several other single transmembrane receptors and proteins can also activate heterotrimeric G proteins, and, therefore, the methodologies described in this chapter can be adapted to other systems.

**Key Words:** Epidermal growth factor (EGF) receptor; GTP-binding proteins; stimulatory GTP-binding protein of adenylyl cyclase ($G_s$); adenylyl cyclase; cAMP.

## 1. Introduction

Ligand binding to the EGF receptor initiates both the activation of mitogenic signal transduction pathways and trafficking events that relocalize the receptor on the cell surface and within intracellular compartments. Upon binding of its ligand, the EGF receptor tyrosine kinase activity is stimulated and the receptor is autophosphorylated on multiple tyrosine residues (1–3). Signaling proteins with Src homology 2 (SH2) domains (4–6) bind to phosphorylated

From: *Methods in Molecular Biology, vol. 327: Epidermal Growth Factor: Methods and Protocols*
Edited by: T. B. Patel and P. J. Bertics © Humana Press Inc., Totowa, NJ

tyrosine residues on the receptor to initiate multiple signaling cascades such as the extracellular regulated kinase (Erk) and phospholipase C$\gamma$.

Studies from our laboratory (7) as well as those from other groups *(8,9)* have shown that in addition to the signaling cascades mentioned above, in several tissues and cells types EGF also activates biological processes and second-messenger systems that are controlled by heterotrimeric G proteins. For instance, in the heart and cardiac myocytes, EGF increases cAMP accumulation and produces inotropic and chronotropic action *(10–14)*. These actions of EGF are mediated by activation of the heterotrimeric stimulatory guanosine 5'-triphosphate (GTP)-binding protein of adenylyl cyclase, $G_s$, and stimulation of the latter enzyme *(10,13,14)*. This action of the EGF receptor is similar to that of the seven transmembrane domain receptors such as the $\beta$-adrenergic receptor that couples with $G_s$. Notably, the EGF receptor is not the only single transmembrane protein that can activate heterotrimeric G proteins. Other examples are the insulin-like growth factor (IGF)-I receptor, insulin receptor, fibroblast growth factor (FGF) receptor, platelet-derived growth factor (PDGF) receptor, and even proteins without tyrosine kinase activity, such as the Alzheimers precursor protein (*see* **ref. 15** for review).

The activation of G proteins as well as the more classical pathways associated with EGF receptor signaling may provide the means to extend the repertoire of biological actions mediated by the growth factor. Moreover, these additional G-protein-mediated signals may provide a mode of fine-tuning other signaling cascades activated by EGF. For instance, the activation of cAMP-dependent protein kinase due to EGF-mediated increase in cAMP levels may determine the extent to which the Erk cascade is activated by this growth factor. It has been demonstrated that c-Raf is inhibited, whereas B-Raf is activated by cAMP-dependent protein kinase, and, therefore, depending upon the Raf isoform expressed, cAMP can either inhibit or augment Erk activation *(16)*.

The recognition that single transmembrane proteins such as the EGF receptor can activate heterotrimeric G proteins led to a search for mechanisms by which these single transmembrane proteins may mimic the actions of classical, heptahelical, G-protein-coupled receptors. To this end, studies from the Nishimoto laboratory which showed that peptides corresponding to short (14- or 15-amino-acid) regions in the IGF-II and $\beta$2-adrenergic receptors could mimic the actions of the activated receptors in terms of stimulating the heterotrimeric G proteins $G_i$ and $G_s$, respectively, provided an important breakthrough *(17–19)*. From these studies, the criteria for regions within receptors that determine the ability to couple with G proteins emerged. These criteria are: (1) the amino-terminus of the region or peptide must contain two basic residues; (2) the carboxy-terminus of the region or peptide must contain the sequence B-B-X-B or B-B-X-X-B (B: basic residues; X: any residue); (3) the sequence length of the region or peptide must be between 14 and 18 residues long. Based upon these criteria, a short sequence within the M$_4$ muscarinic cholinergic receptor was identified to be responsible for the coupling with, and activation of, G$_i$ *(20)*. Moreover, because the EGF receptor can stimulate adenylyl cyclase via activation of G$_s$, we investigated if the intracellular segment of the EGF receptor contained motifs that could activate G proteins. Using the criteria described above, we identified two regions in the EGF receptor encompassed by amino acids 645–657 (EGFR-13) and amino acids 680–693 (EGFR-14) that were candidates as G$_s$ activators. In this report, using peptides corresponding to regions in the EGF receptor, we have presented three different approaches by which activation of the trimeric Gs can be monitored. Additionally, using the full-length EGF receptor and recombinant Gα$_s$, we have provided the protocol to determine the stoichiometry of EGF receptor association with the G protein.

## 2. Materials

The following methods require protein purification equipment, a scintillation counter, experience in biochemical assays, and polyacrylamide gel electrophoresis and Western blotting apparatus.

### 2.1. Monitoring the Binding of GTPγS to Heterotrimeric G$_s$

1. Purified, recombinant Gα$_s$ expressed in *Escherichia coli*.
2. Purified βγ subunits from bovine brain.
3. Synthetic peptides EGFR 13 (RRREIVRKRTLRR), phospho-EGFR 13 (same sequence as EGFR 13 except that Thr residue is phosphorylated), and EGFR-14 (HLRILKETEFKKIK) (*see* **Notes 1** and **2**).
4. [$^{35}$S]GTPγS.
5. Unlabeled GTPγS.
6. Nitrocellulose filters (BA85, 25-mm circles) from Scheilcher & Schuell.
7. Ethylene glycol monoethyl ether
8. Binding buffer: 50 m*M* hydroxyethyl piperazine ethan sulfonate (HEPES), pH 8.0, 120 μ*M* MgCl$_2$, 100 μ*M* ethylenediaminetetraacetic acid (EDTA), 1 m*M* dithiothreitol (DTT).
9. Stopping buffer: 25 m*M* Tris-HCl, pH 7.4, 10 m*M* NaCl, 25 m*M* MgCl$_2$ (*see* **Note 3**).
10. Vacuum assisted filtration system for 25-mm diameter filters such as that made by Millipore.
11. Scintillation counter.

### 2.2. Steady-State GTPase Activity Measurements

In addition to **Subheading 2.1.**, **items 1–3**, the following will be required.

1. Norit A Charcoal.
2. [γ-³²P]GTP.
3. 50 m$M$ NaH$_2$PO$_4$ (pH 8.0).
4. Assay medium: 25 m$M$ HEPES, pH 8.0, 110 μ$M$ EDTA, 200 μ$M$ MgSO$_4$, 1 m$M$ DTT.
5. Centrifuge for Eppendorf tubes.
6. Scintillation counter.

## 2.3. Adenylyl Cyclase Activity Assays With S49 cyc⁻ Cell Membranes

In addition to **Subheading 2.1.**, **items 1–3**:

1. S49 cyc⁻ cell membranes.
2. Guanosine triphosphate (GTP)γS.
3. Guanosine diphosphate (GDP)βS.
4. [α-³²P]ATP.
5. [³H]cAMP.
6. Dry ice.
7. Dowex 50 (AG50-X8) columns comprising 1 mL of Dowex AG50 resin in 10-mL disposable plastic column from Biorad.
8. Alumina columns consisting of 0.3 g of alumina (Fisher Scientific) in 10-mL disposable columns from Biorad.
9. Hydrochloric acid, 0.1 $N$.
10. Imidazole, 0.1 $M$.
11. Assay medium: 50 m$M$ HEPES, pH 7.4, 1 m$M$ MgSO$_4$, 100 μ$M$ EDTA, 1 m$M$ isobutylmethylxanthine (IBMX), 1 m$M$ cAMP, 0.1 mg/mL creatine kinase, 12 m$M$ creatine phosphate, 0.1 mg/mL bovine serum albumin (BSA), 10 U/mL myo-kinase and 0.1 m$M$ [α-³²P]-ATP.
12. 2X Stop buffer: 100 m$M$ HEPES, pH 8.0, 2 m$M$ adenosine triphosphate (ATP), 2 m$M$ cAMP, 4% sodium dodecyl sulfate (SDS).

## 2.4. Determining the Stoichiometry of Association Between the EGF Receptor and Gα$_s$

1. Purified EGF receptor (Promega, Madison, WI).
2. Purified Gα$_s$.
3. [α-³²P]ATP.
4. Phosphorylation buffer: 5 m$M$ HEPES, pH 7.4, 5 m$M$ MgCl$_2$, 2 m$M$ MnCl$_2$, 20 μg/mL each of leupeptin and aprotinin, 1 m$M$ DTT, 10 μ$M$ ATP, 1 μ$M$ EGF.
5. Anti-EGF receptor antibody (EGFR1 from Amersham).
6. Immunoprecipitation buffer: 25 m$M$ Tris-HCl, pH 7.4, 0.5% NP-40, 1% Triton X-100, 1 m$M$ EDTA, 150 m$M$ NaCl, 20 μg/mL each of aprotinin and leupeptin.
7. X-ray film developer or phosphor-imager.
8. Scintillation counter.

# 3. Methods

## 3.1. Purification of Gα$_s$ and Gβγ Subunits

Purify the α-subunit of G$_s$ according to the method in **ref. 21**. The Gα$_s$ is expressed in *E. coli* (strain BL21/DE3) transformed with pQE-6 vector containing the short form (45 kDa) Gα$_s$ cDNA inserted downstream of Lac Z gene promoter. Express the protein overnight with IPTG (isopropyl-β-D-thio-galactopyranoside, 0.05 m*M*) at 20°C. Quantify the amount of active Gα$_s$ by performing a binding assay with GTPγS as described below with the modifications in **Note 5**. Typically, between 20 and 50% of the purified protein is active. Store the purified protein at a concentration of ≥1 mg/mL in 20% glycerol.

Purify the βγ subunits from the membrane protein extracts of fresh bovine brain as described in **refs. 22** and **23**. The sodium cholate used should be purified as described in **Note 4**. Pharmacia fast-protein liquid chromatography (FPLC) equipment or its equivalent is used throughout purification.

## 3.2. Binding Studies Using Guanosine 5'-(γ-[³⁵S] thio) Triphosphate

1. Reconstitute heterotrimeric G$_s$ by incubating 1 pmol recombinant Gα$_s$ with bovine brain βγ subunits at a molar concentration of 1:5 on ice for 1 h. Maintain the volume of this mix at about 10 μL. Depending upon the quality of the Gβγ protein preparation, a lower Gα$_s$:Gβγ ratio (1:3) may also be used. *See* **Note 6** on how to optimize the ratio.
2. Incubate the reconstituted G$_s$ on ice with or without the individual peptides corresponding to the EGF receptor for 5 min in binding buffer.
3. Initiate the binding of GTPβγS binding by the addition of 100 n*M* radiolabeled nucleotide in binding buffer as described under **Subheading 2.1.** The final volume of the binding assay is 100 μL.
4. To determine the total radioactivity of GTPγ³⁵S in the assay, transfer 10 μL of binding mix onto the filter.
5. Terminate the binding reaction by transferring into 2 mL of ice-cold stop buffer (*see* **Note 3**).
6. To determine the specific binding of GTPγ³⁵S, reactions similar to those in **steps 1–4**, are performed with the same amount of radioactive GTPγ³⁵S but 100 μ*M* of the unlabeled nucleotide.
7. Separate the bound GTPγS from the free nucleotide by filtering the stopped reaction mixture through Schleicher and Schull BA 85 nitrocellulose filters (under vacuum).
8. Wash the filters from **step 7** three times with 2 mL each of ice-cold stopping buffer, place on a Whatman filter paper, and allow to air-dry.
9. Transfer the dried filters in to scintillation vials and dissolve in 2 mL of ethylene glycol mono-ethyl-ether.
10. Monitor the amount of radioactivity associated with the filters using a scintillation counter.

a. Protocol 1: To study the time course of GTPγ35S binding to $G_s$, incubate reconstituted $G_s$ in 1 mL of assay buffer in the presence or absence of peptides for 5 min. Initiate binding by addition of 100 n$M$ of GTPγ$^{35}$S. At various time points (e.g., 0, 10, 15, 20, 30 min, and so on up to 1 h) withdraw 100-μL aliquots of the binding reaction and mix with 2 mL of ice-cold stopping buffer. Keep the samples on ice until all the time points are collected. Determine bound GTPγ$^{35}$S as described previously. For determining GTPγ$^{35}$S binding at zero time, add 90 μL of binding buffer containing GTPγ$^{35}$S and 10 μL Gs directly to ice-cold stopping buffer.

b. Protocol 2: To study the dose-response relationship, incubate peptides of various concentrations with the reconstituted $G_s$ in 100 μL of assay buffer. Allow the binding reaction to proceed for 1 h at room temperature. Terminate the reaction by adding 2 mL of ice-cold stopping buffer. Determine bound radionucleotide as described previously.

10. Subtract the nonspecific binding of GTPγS (**step 6**) from the total binding (**step 5**) divided by the specific radioactivity of the GTPγS determined from **step 4**. This will yield pmol of GTPγS bound to $G\alpha_s$ in 90 μL of the reaction mix. Dividing this number by the number of pmol of $G\alpha_s$ in the 90 μL of reaction mix will yield pmol nucleotide bound to pmol G protein.

### 3.3. Steady-State GTPase Activity Measurements

1. To monitor the steady-state GTPase activity at various times, incubate reconstituted $G_s$ proteins (90 μL) with or without peptides (90 μL) at 25°C in 820 μL of medium containing the following at final concentration: 25 m$M$ HEPES, pH 8.0, 110 μ$M$ EDTA, 200 μ$M$ $MgSO_4$, and 1 m$M$ DTT.

2. Initiate the reaction by adding 100 n$M$ [γ-$^{32}$P]GTP.

3. Keep a small aliquot (10 μ/L) of the reaction mixture to determine total radioactivity and specific radioactivity of the [γ-$^{32}$P]GTP.

4. At various time points withdraw 100-μL aliquots and mix by vortexing with 0.75 mL of ice-cold 5% (w/v) Norit A in 50 m$M$ $NaH_2PO_4$, pH 8.0 (*see* **Note 7**).

5. Keep the mixture on ice until all the time points are collected.

6. To determine the release of radiolabeled ortho-phosphate at zero time point, add 90 μL of assay medium containing GTPγ$^{32}$P into Norit A before starting the assay.

7. Centrifuge the mix at 14,000$g$ for 10 min. Remove supernatant and subject it to a second round of centrifugation.

8. Transfer an aliquot (0.65 mL) of the supernatant and determine $^{32}$P-labeled phosphate contents by scintillation counting.

a. Protocol 1: For dose–response studies, incubate reconstituted $G_s$ proteins (10 μL) with or without the peptides (10 μL) with different concentration at 25°C in 100 μL of medium. Because the rates of GTPase activity were found to be linear for at least 20 min of assay under all experimental conditions, this assay can be conducted for 10 min.

9. From the total amount of $^{32}$P-labeled *ortho*-phosphate released and specific radioactivity of the [γ-$^{32}$P]GTP (*see* **step 3**), the steady-state GTPase activity over 10 min/pmol of $G\alpha_s$ can be calculated.

### 3.4. Adenylyl Cyclase Assays With S49 cyc⁻ Cell Membranes

In this assay the functional activity of G$_s$ is measured by the ability of the activated G$_s$ to stimulate adenylyl cyclase activity in mouse lymphoma S49 cyc⁻ cell membranes. S49 cyc⁻ cells do not express endogenous Gα$_s$ and, therefore, present a valuable system in which the activity of exogenously added G protein can be monitored. The cell membranes were isolated as described by Iyengar et al. *(24)*.

1. For adenylyl cyclase activity assays, in a final volume of 100 μL, incubate reconstituted Gs (10 n*M*) in the presence or absence of peptides with GTPγS (100 n*M*) for 1 h as described in GTPγS binding protocol (*see* **Subheading 3.2.**).
2. Thereafter, transfer 0.1 pmol (10 μL of the binding reaction) of G$_s$ to 10 μg of S49 cyc⁻ cell membrane protein in the presence of 1 nmol of GDPβS. At this point the incubation volume is 20 μL.
3. Initiate adenylyl cyclase assays (100 μL final volume) by adding 80 μL of assay medium (*see* **Note 8**).
4. Let the assay proceed for 15 min at room temperature with continuous agitation (*see* **Note 9**).
5. Terminate the reaction by adding 100 μL of 2X stop buffer containing 100 m*M* HEPES, pH 8.0, 2 m*M* ATP, 2 m*M* cAMP, and 4% SDS followed by chilling on dry ice.
6. Separate the reaction product $^{32}$P cAMP from unused substrate α-$^{32}$P ATP by sequential column chromatography with Dowex 50 column followed by alumina column as described in **ref. 25** (*see* **Note 10**).
7. Determination of recovery of cAMP after the final column chromatography is calculated from the percentage of $^3$H cAMP eluted from the columns. The $^{32}$P dpm in the cAMP fraction is corrected for the recovery and then converted to pmole cAMP formed using the specific radioactivity of [α-$^{32}$P]ATP. After dividing by the time of the assay (15 min) and the protein content of S49 cyc⁻ cells, the data are finally expressed as pmol cAMP/mg protein/min.

### 3.5. Stoichiometry of Association Between the EGF Receptor and Gα$_s$

This method relies on our previous finding that the EGF receptor can stoichiometrically phosphorylate the α subunit G$_s$ *(7)*. Essentially, after stoichiometric phosphorylation, the EGF receptor/G protein complex is immunoprecipiated and the amount of G protein associated with the receptor calculated.

1. Incubate 50 ng (~900 fmol) of Gα$_s$ with 66 ng of purified EGF receptor in a final volume of 20 μL with the phosphorylation buffer described under **Subheading 2.4.** Supplement the buffer with [γ-$^{32}$P]ATP so that the specific radioactivity is approx 1000 dpm/pmol.
2. Incubate the mixture for 1 h at room temperature.
3. Transfer 10 μL of the reaction mixture to 500 μL of immunoprecipitation buffer.
4. To the other 10 μL of the phosphorylation mix from **step 2**, add 20 μL of 1.5X Laemmli sample medium and boil for 5 min.

5. To the mixture described in **step 3**, add 2–3 µg/mL of EGFR1 antibody or the anti-Gα$_s$ antibody (UBI).
6. Control immunoprecipitates like those in **step 4** should be conducted with anti-mouse IgG or anti-rabbit IgG instead of the EGFR1 and anti-Gα$_s$ antibody, respectively.
7. Incubate the mixture with antibodies overnight at 4°C with constant rolling.
8. Add 20 µL of pansorbin suspension (10%, Calbiochem, CA; *see* **Note 11**) and incubate further at 4°C for 1 h.
9. Centrifuge the mixture at 14,000g for 1 min and then wash the pellet with high-salt buffer described under **Subheading 2.4.**
10. Repeat step 8, except wash with no-salt buffer as described under **Subheading 2.4.**
11. Resuspend the final pellet in 30 µL of Laemmli sample medium, boil sample for 5 min, and centrifuge at 14,000g for 1 min.
12. Subject the proteins in the Laemmli sample medium from **steps 4** and **11**, to SDS-PAGE using 10% gels.
13. After staining and fixing proteins, dry the gel and subject to autoradiography (*see* **Note 12**).
14. Lay the gel over the autoradiograph and mark the positions of the phosphorylated EGF receptor and Gα$_s$.
15. Cut out the bands corresponding to the EGF receptor and Gα$_s$ and determine the amount of $^{32}$P label associated with each.
16. For the samples derived from **step 4**, divide the amount of radioactivity associated with the EGF receptor and Gα$_s$ bands by the proteins in these bands EGF receptor (33 ng) and Gα$_s$ bands (25 ng). This will provide the dpm of $^{32}$P label/pmol of each protein. From the specific radioactivity of [γ-$^{32}$P]ATP, one can work out the pmol of phosphate incorporated. Typically, we have found this to be around 5 mol Pi/mol EGF receptor and 2 mol Pi/mol Gα$_s$.
17. In the immunoprecipitated samples, divide the dpm associated with the EGF receptor and Gα$_s$ by the dpm/pmol of phosphate incorporated in each protein (**step 16**; also *see* **Note 13**). This will provide the pmol of each protein associated with each other. Our experience is that 1 mol of the receptor associates with 1 mol of Gα$_s$.

## 4. Notes

1. The peptides should be as pure as possible (>95%) to avoid artifacts. The phosphothreonine in the phospho-EGFR 13 peptide corresponds to the protein kinase C (PKC) phosphorylation site (Thr 654) on the EGFR.
2. Previous evidence from our lab had suggested that PKC-mediated phosphorylation of the EGF receptor obliterates the ability of EGF to activate adenylyl cyclase activity *(7,26)*. Therefore, the phospho-EGFR 13 is a good control.
3. For each sample, a 2-mL aliquots (in a 5-mL tube) of the stop buffer should be kept on a slush of ice.
4. Purify the sodium cholate by passing a 20% w/v solution over a 2-(diethylamino)-ethylamine (DEAE) column (60 cm long × 5 cm diameter) several times and then

precipitate the sodium cholate by decreasing pH. Filter and wash the powder with ether and dry.

5. The maximal and stoichiometric binding of Gα$_s$ to GTPγS in a 100 µL volume is assessed at a final concentration of Gα$_s$ 10 nM in the presence of 1 µ$M$ GTPγ$^{35}$S and 25 m$M$ MgCl$_2$ over a 60-min period at room temperature; the time may have to be empirically determined for saturation of binding. Assuming that each mole of GTPγS binds 1 mol of Gα$_s$ and the total molar concentration of G$_s$ based in protein content, the percent of active Gα$_s$ can be estimated.

6. The steady-state GTPase activity of heterotrimeric G$_s$ is significantly lower than that of the monomeric θ subunit. Therefore, one can titrate in the amount of Gβγ in the reconstitution of G$_s$ and determine the amount of Gβγ that maximally decreases the steady-state GTPase activity.

7. The Norit A (charcoal) suspension should be aliquoted into stop tubes while it is stirred continuosly. This ensures an equal amount of charcoal in each tube. If the background readings (blanks without G protein) are high, then spin the charcoal suspension at 20,000$g$ and remove the supernatant (this may have to be repeated to make sure that very small charcoal particles are removed). Add back the same volume of sodium phosphate that you remove.

8. [$^3$H] cAMP is added in the assay mixture as an internal standard to determine cAMP recovery after the column chromatography. Therefore, it is essential to determine the amount of [$^3$H]cAMP as well as the specific radioactivity of the [α-$^{32}$P]ATP.

9. The best method of continuous agitation for small assays (100 µL) that we have found is to place the sealed Eppendorf tubes into holes on a circular foam platform that is attached to a vortex mixer. The platform adapters are commercially available.

10. Dowex 50 (AG50-X8) columns were washed with 3 mL of 0.1 $N$ HCl and 30 mL of deionized water. Thaw samples (200 µL each) and apply 180 µL onto 3 mL Dowex 50 columns. Wash the columns with 3 mL of deionized water. The eluate from this wash mainly contains unused [α-$^{32}$P]ATP and can be discarded. Then place the Dowex column over the alumina column (neutral alumina, 50–200 mesh) so that eluate drips directly on to the alumina. Wash the Dowex column with 9 mL of deionized water. cAMP eluted from the Dowex column is retained on the alumina column. Alumina columns are then placed over scintillation vials and eluted with 7 mL of 0.1 $M$ imidazole, pH 7.4. The reaction product, cAMP, eluted in this step is collected in scintillation vials; after adding scintillation fluid, monitor the $^{32}$P and $^3$H contents using a scintillation counter.

11. The pansorbin suspension should be washed with immunoprecipitation buffer; then resuspend the pansorbin in nine times the volume of the immunoprecipitation buffer prior to use.

12. To decrease the background radioactivity in the gel from free phosphate or left over [γ-$^{32}$P]ATP, wash the gel extensively in fixing buffer.

13. For the EGF receptor and Gα$_s$ bands in the samples from immunoprecipitations, make sure to subtract the radioactivity corresponding to similar positions on the gel in the control immunoprecipitations performed with nonspecific IgG.

## References

1. Hunter, T. and Cooper, J. A. (1981) Epidermal growth factor induces rapid tyrosine phosphorylation of proteins in A431 human tumor cells. *Cell* **24,** 741–752.
2. Carpenter, G. (1987) Receptors for epidermal growth factor and other polypeptide mitogens. *Annu. Rev. Biochem.* **56,** 881–914.
3. Hunter, T. and Cooper, J. A. (1985) Protein-tyrosine kinases. *Annu. Rev. Biochem.* **54,** 897–930.
4. Cantley, L. C., Auger, K. R., Carpenter, C., et al. (1991) Oncogenes and signal transduction. *Cell* **64,** 281–302.
5. Pawson, T. and Gish, G. D. (1992) SH2 and SH3 domains: from structure to function. *Cell* **71,** 359–362.
6. Satoh, T., Nakafuku, M., and Kaziro, Y. (1992) Function of Ras as a molecular switch in signal transduction. *J. Biol. Chem.* **267,** 24,149–24,152.
7. Poppleton, H., Sun, H., Fulgham, D., Bertics, P., and Patel, T. B. (1996) Activation of Gsalpha by the epidermal growth factor receptor involves phosphorylation. *J. Biol. Chem.* **271,** 6947–6951.
8. Carpenter, G. and Cohen, S. (1990) Epidermal growth factor. *J. Biol. Chem.* **265,** 7709–7712.
9. Carpenter, G. and Wahl, M. I. (1990) Peptide growth factors and their receptors. *Handb. Exp. Pharmacol.* **95,** 169–171.
10. Nair, B. G., Rashed, H. M., and Patel, T. B. (1993) Epidermal growth factor produces inotropic and chronotropic effects in rat hearts by increasing cyclic AMP accumulation. *Growth Factors* **8,** 41–48.
11. Yu, Y., Nair, B. G., and Patel, T. B. (1992) Epidermal growth factor stimulates cAMP accumulation in cultured rat cardiac myocytes. *J. Cell Physiol.* **150,** 559–567.
12. Barbier, A. J., Poppleton, H. M., Yigzaw, Y., et al. (1999) Transmodulation of epidermal growth factor receptor function by cyclic AMP-dependent protein kinase. *J. Biol. Chem.* **274,** 14,067–14,073.
13. Nair, B. G., Rashed, H. M., and Patel, T. B. (1989) Epidermal growth factor stimulates rat cardiac adenylate cyclase through a GTP-binding regulatory protein. *Biochem. J.* **264,** 563–571.
14. Nair, B. G., Parikh, B., Milligan, G., and Patel, T. B. (1990) Gs alpha mediates epidermal growth factor-elicited stimulation of rat cardiac adenylate cyclase. *J. Biol. Chem.* **265,** 21,317–21,322.
15. Patel, T. B. (2004) Single transmembrane spanning heterotrimeric g protein-coupled receptors and their signaling cascades. *Pharmacol. Rev.* **56,** 371–385.
16. Rudolph, J. A., Poccia, J. L., and Cohen, M. B. (2004) Cyclic AMP Activation of the Extracellular Signal-regulated Kinases 1 and 2: implications for intestinal cell survival throught the transient inhibition of apoptosis. *J. Biol. Chem.* **279,** 14,828–14,834.
17. Nishimoto, I., Murayama, Y., Katada, T., Ui, M., and Ogata, E. (1989) Possible direct linkage of insulin-like growth factor-II receptor with guanine nucleotide-binding proteins. *J. Biol. Chem.* **264,** 14,029–14,038.

18. Okamoto, T., Katada, T., Murayama, Y., Ui, M., Ogata, E., and Nishimoto, I. (1990) A simple structure encodes G protein-activating function of the IGF-II/ mannose 6-phosphate receptor. *Cell* **62,** 709–717.

19. Okamoto, T., Murayama, Y., Hayashi, Y., Inagaki, M., Ogata, E., and Nishimoto, I. (1991) Identification of a Gs activator region of the beta 2-adrenergic receptor that is autoregulated via protein kinase A-dependent phosphorylation. *Cell* **67,** 723–730.

20. Okamoto, T., and Nishimoto, I. (1992) Detection of G protein-activator regions in M4 subtype muscarinic, cholinergic, and alpha 2-adrenergic receptors based upon characteristics in primary structure. *J. Biol. Chem.* **267,** 8342–8346.

21. Graziano MP, F. M. a. G. A. (1989) Expression of Gsα in *Escherichia coli*. *J. Biol. Chem.* **264,** 409–418.

22. Mumby, S., Pang, I. H., Gilman, A. G., and Sternweis, P. C. (1988) Chromatographic resolution and immunologic identification of the alpha 40 and alpha 41 subunits of guanine nucleotide-binding regulatory proteins from bovine brain. *J. Biol. Chem.* **263,** 2020–2026.

23. Neer, E. J., Lok, J. M., and Wolf, L. G. (1984) Purification and properties of the inhibitory guanine nucleotide regulatory unit of brain adenylate cyclase. *J. Biol. Chem.* **259,** 14,222–14,229.

24. Iyengar, R., Bhat, M. K., Riser, M. E., and Birnbaumer, L. (1981) Receptor-specific desensitization of the S49 lymphoma cell adenylyl cyclase. Unaltered behavior of the regulatory component. *J. Biol. Chem.* **256,** 4810–4815.

25. Salomon, Y., Londos, C., and Rodbell, M. (1974) A highly sensitive adenylate cyclase assay. *Anal. Biochem.* **58,** 541–548.

26. Poppleton, H. M., Wiepz, G. J., Bertics, P. J., and Patel, T. B. (1999) Modulation of the protein tyrosine kinase activity and autophosphorylation of the epidermal growth factor receptor by its juxtamembrane region. *Arch. Biochem. Biophys.* **363,** 227–236.

# 5

# Regulator of Epidermal Growth Factor Signaling

*Sprouty*

## Esther Sook Miin Wong and Graeme R. Guy

## Summary

Sprouty was first discovered through its downregulatory effect on fibroblast growth factor (FGF) receptor pathway during tracheal development. Sprouty expression is also induced by the epidermal growth factor receptor (EGFR) cascade in some tissues, including the follicle cells of the ovary, the wing, and eye imaginal disc, and acts to abolish mitogen-activated protein (MAP) kinase activated by EGFR signaling. Sprouty is an intracellular protein that translocates to membrane ruffles upon EGF stimulation by virtue of a translocatory domain within its highly conserved cysteine-rich C-terminal region. Human Sprouty2 (hSpry2) binds the catalytic RING Finger of Casitas B-lineage lymphoma (c-Cbl), an E3 ubiquitin ligase that has been identified to target EGFR degradation. Overexpressed hSpry2 induces a prolonged EGFR-mediated MAP kinase activation. hSpry2 acts to sequester c-Cbl molecules from activated EGFR and impedes EGFR ubiquitination and downregulation, thereby potentiating the amplitude and longevity of intracellular signals. The strategies described herein encompass various methods that have been used to measure the status of EGFR following ectopic expression of hSpry2.

**Key Words:** Sprouty; c-Cbl; EGFR; MAP kinase; downregulation; ubiquitination; endocytosis.

## 1. Introduction

Growth factors and their cognate receptor tyrosine kinases (RTKs) play pivotal roles in embryonic development, signaling in mature cells, and regulation of pathogenesis. The mitogen-activated protein (MAP) kinase cascade is a central intracellular signaling pathway linking activation of surface receptors to cytoplasmic and nuclear effectors by transducing signals from RTKs to two members of the Ras family of small G proteins, Ras and Rap1 (*1*). Ras interacts directly with and activates the Raf serine/threonine kinases (c-Raf and B-Raf),

From: *Methods in Molecular Biology, vol. 327: Epidermal Growth Factor: Methods and Protocols*
Edited by: T. B. Patel and P. J. Bertics © Humana Press Inc., Totowa, NJ

phosphorylated Raf then activates MEK, which in turn phosphorylates and activates the MAP kinases (ERK1 and ERK2) by double phosphorylation on threonine and tyrosine residues. ERK1/2 have been shown to be essential for cellular proliferation as well as phenotypic determination (2). Various regulatory proteins and feedback mechanisms tightly control these key pathways.

Post-RTK signal transduction is regulated primarily by receptor endocytosis and degradation in lysosomes ("receptor downregulation"). The ErbB family of RTKs exemplifies the importance of stringent regulation on signaling potency and its impact on the formation and malignant progression of many human cancers (3). Once triggered by ligands, phosphorylation occurs on the intracellular tyrosine kinase domain, and receptor activation leads to the recruitment and phosphorylation of several downstream intracellular substrates, mitogenic signaling, and other growth-promoting cellular activities. Of the four ErbB family members, ErbB-1 (or EGFR) uniquely binds to c-Cbl, and is effectively targeted for lysosomal/proteasomal degradation (4). Ubiquitination of proteins constitutes an important cellular mechanism for targeting short-lived proteins for degradation; it is a posttranslational modification in which ubiquitin chains or single ubiquitin molecules are appended to target proteins, giving rise to poly- or monoubiquitination, respectively (5,6). c-Cbl functions as an E3 ubiquitin ligase that docks onto phosphorylated tyrosine residue 1045 ($Y^{1045}$) on EGFR, bringing them in close proximity to E2 ubiquitin-conjugating enzymes for the relay of activated ubiquitin moieties, thus marking receptors for lysosomal/proteasomal destruction and consequent signal attenuation (7).

*Drosophila* Sprouty (dSpry) was identified in a genetic screen as an inhibitor of EGFR-triggered cell recruitment in the eye, which is transcriptionally induced by the pathway (8). dSpry localizes to the inside surface of the plasma membrane and intercepts signaling at a point between the receptor and activation of Ras. Presently, four mammalian Spry proteins (Spry1–4) have been cloned and sequenced (9,10). Spry proteins have distinctive cysteine-rich C-terminal regions flanked by cysteine-free regions that lack any recognizable protein–protein interaction domains. A short sequence in the N-termini of Spry proteins and several pockets of residues distributed throughout the protein family are also conserved. The human homolog Sprouty2 (hSpry2), is a 35-kDa polypeptide bearing 51% amino acid identity to dSpry. An in vivo candidate approach first employed to identify any role of hSpry2 in EGFR signaling revealed that hSpry2 binds c-Cbl through a divergent, protein-interacting N-terminal region of hSpry2 (11), while the conserved cysteine-rich domain targets the protein to membrane ruffles upon EGF stimulation (12). This chapter describes a series of methodologies employed to follow-up from the initial identification of a hSpry2-c-Cbl interaction leading to elucidation of the functional consequence of hSpry2-c-Cbl binding on the regulation of MAP kinase signaling by EGFR *(13)*. Specifically, this chapter will discuss methods for assessing the disposition of ligand-bound receptors, the kinetics of EGFR downregulation, ubiquitination, and endocytosis, and physiological influence on the differentiation potential of PC12 cells. For additional methods related to EGFR ubiquitination and degradation, *see* Chapters 8 and 9.

## 2. Materials

All molecular methods highlighted in this chapter require equipment and expertise in DNA cloning, polymerase chain reaction (PCR), sodium dodecyl sulfate-polyacrylamide gel electrophoresis (SDS-PAGE), Western blotting, immunofluorescence, and mammalian cell culture techniques.

### 2.1. Full-Length Sequences and Vectors

1. hSpry2 (GenBank™ accession number AF039843).
2. c-Cbl (X57110).
3. EGFR (X57110).
4. pGEX4T1 vector for bacterial production of glutathione-*S*-transferase (GST)-tagged fusion proteins.
5. pXJ40HA, pXJ40FLAG, pMyc mammalian expression vectors.
6. pEGFP-C3 vector (CLONTECH Laboratories, Inc., Palo Alto, CA).

### 2.2. Confocal Microscopy and Immunofluorescence Analyses

1. COS-1 monkey kidney cells.
2. Dulbecco's modified Eagle's medium (DMEM) complete growth medium: supplemented with 10 m$M$ hydroxyethyl piperazine ethane sulfonate (HEPES) (pH 7.4), 2 m$M$ glutamine, 100 U/mL penicillin/streptomycin, and 10% fetal bovine serum (FBS, HyCloneLaboratories™, Inc., Logan, UT).
3. LipofectAMINE™ reagent (Life Technologies, Inc., Carlsbad, CA).
4. Human recombinant EGF (Upstate Biotechnology, Lake Placid, NY).
5. Phosphate-buffered saline (PBS): 140 m$M$ NaCl, 2.7 m$M$ KCl, 10 m$M$ Na$_2$HPO$_4$, 1.8 m$M$ KH$_2$PO$_4$, dissolve in water, adjust to pH 7.4 with HCl.
6. PBSCM: PBS containing 10 m$M$ CaCl$_2$, 10 m$M$ MgCl$_2$.
7. 3% paraformaldehyde (Sigma Chemicals, St. Louis, MO) in PBSCM.
8. 0.1% saponin (cat. no. S-7900, Sigma).
9. Fluorescence dilution buffer FDB: 5% FBS, 2% bovine serum albumin (BSA) in PBSCM.
10. For primary antibody staining: anti-c-Cbl (C-15) polyclonal (Santa Cruz Biotechnology, Santa Cruz, CA); anti-EGFR (E12020) monoclonal (Transduction Laboratories, Lexington, KY); anti-FLAG polyclonal (Affinity Bioreagents, Inc., Golden, CO).
11. For secondary antibody staining: Fluorescein isothiocyanate (FITC)-conjugated AffiniPure rabbit anti-mouse IgG; Texas Red® dye-conjugated AffiniPure goat anti-rabbit IgG (Jackson ImmunoResearch Laboratories, Inc., West Grove, PA).

12. Crystal mount reagent (BiØmeda Co., Foster City, CA).
13. Bio-Rad MRC-1024 confocal laser optics attached to a microscope (Zeiss, Oberkochen, Germany) interfaced with an argon/krypton laser; LaserSharp software (Bio-Rad).

## 2.3. Reagents/Special Equipment for Evaluating the Kinetics of EGFR Downregulation

1. COS-1 monkey kidney cells.
2. LipofectAMINE™ reagent (Life Technologies, Inc., Carlsbad, CA).
3. Human recombinant EGF (Upstate Biotechnology, Lake Placid, NY).
4. Human recombinant $^{125}$I-EGF (Amersham Pharmacia Biotech, Buckinghamshire, HP).
5. MG132 prepared in dimethyl sulfoxide (DMSO) (Sigma).
6. Chloroquine diphosphate salt (cat. no. C-6628, Sigma); use at 25 m$M$ in PBS.
7. Binding buffer: DMEM containing 0.5% BSA, 20 m$M$ HEPES.
8. Ligand stripping buffer: 150 m$M$ acetic acid, 150 m$M$ NaCl, pH 2.7.
9. Solubilization buffer: 0.1 $M$ NaOH, 0.1% SDS.
10. Monensin sodium salt (BIOMOL Research Laboratories, Inc., Plymouth, PA).
11. 4β-Phorbol 12-myristate 13-acetate (PMA; Calbiochem, San Diego, CA).
12. LKB-Wallac 1282 CompuGamma counter.

## 2.4. Measuring Cell Surface Antigen Staining

1. COS-1 monkey kidney cells.
2. LipofectAMINE™ reagent (Life Technologies, Inc., Carlsbad, CA).
3. Human recombinant EGF (Upstate Biotechnology, Lake Placid, NY).
4. Assay buffer: PBS, 3% FBS, 0.1% sodium azide, pH 7.4.
5. EGFR mouse monoclonal IgG$_{2a}$ (clone 528, Santa Cruz Biotech, Santa Cruz, CA).
6. Texas red-conjugated goat anti-mouse IgG (Jackson ImmunoResearch Labs, West Grove, PA).
7. Propidium iodide solution, 10 µg/mL in PBS, storage at 4°C (Sigma).
8. Flow cytometer (Beckton-Dickinson, Franklin Lakes, NJ).

## 2.5. Ascertaining EGFR Ubiquitination In Vivo

1. Human kidney epithelial 293T cells or Chinese hamster ovary (CHO) cells.
2. RPMI 1640 complete growth medium: supplemented with 10 m$M$ HEPES (pH 7.4), 2 m$M$ glutamine, 100 U/mL penicillin/streptomycin, and 10% FBS (HyCloneLaboratories®, Logan, UT).
3. LipofectAMINE™ reagent (Life Technologies, Inc., Carlsbad, CA).
4. Human recombinant EGF (Upstate Biotechnology, Lake Placid, NY).
5. RIPA buffer: 50 m$M$ Tris-HCl (pH 7.4), 150 m$M$ NaCl, 0.25 m$M$ EDTA (pH 8.0), 1% Triton X-100, 1% sodium deoxycholate, 0.2% sodium fluoride, 0.1% orthovanadate. Add 1X protease inhibitor cocktail (Boehringer Mannheim) to a

final concentration of: 1 m$M$ phenylmethylsulfonyl fluoride (PMSF), 0.5 μ$M$ sodium fluoride, 5 μg/mL aprotinin, 5 μg/mL leupeptin, 1 μg/mL pepstatin A, and 0.1% sodium orthovanadate (phosphatase inhibitor, Sigma) before use.

6. Bio-Rad Protein Assay (Bio-Rad, Hercules, CA).
7. EGFR polyclonal conjugated to agarose (AC beads) (500 μg IgG/0.25 mL agarose) (cat. no. 1005, Santa Cruz Biotech, Santa Cruz, CA).
8. 2X Laemmli SDS-PAGE sample buffer containing 2-mercaptoethanol (added freshly).
9. SDS-PAGE equipment.
10. Blocking buffer: PBS containing 1% BSA plus 0.1% Tween 20.
11. Stripping buffer: 62.5 m$M$ Tris-HCl (pH 6.8), 2% SDS.
12. Anti-HA and anti-FLAG monoclonal antibodies (Roche Molecular Biochemicals, Indianapolis, IN).
13. Anti-EGFR monoclonal antibody (cat. no. E12020, Transduction Laboratories, Lexington, KY).
14. Secondary rabbit anti-mouse antibody conjugated to horseradish peroxidase (HRP) (Sigma).
15. Enhanced chemiluminescence (ECL) reagent (Amersham Pharmacia Biotech, Buckinghamshire, HP).

## 2.6. Assaying EGFR Ubiquitination In Vitro

1. Bacterial glutathione-*S*-transferase (GST) expression constructs.
2. Glutathione Sepharose® 4B (Pharmacia Biotech, Uppsala, AB).
3. Elution buffer: 10 m$M$ reduced glutathione in 50 m$M$ Tris-HCL, pH 8.0.
4. COS-1 monkey kidney cells or A431 human epidermoid carcinoma cells.
5. EGFR polyclonal conjugated to agarose (AC beads) (500 μg IgG/0.25 mL agarose) (cat. no. 1005, Santa Cruz Biotech, Santa Cruz, CA).
6. 1X Reaction buffer: 50 m$M$ Tris-HCl (pH 7.4), 2.5 m$M$ MgCl, 2 m$M$ DTT, 2 m$M$ ATP.
7. Mammalian ubiquitin (AFFINITI Research, Mamhead, EX).
8. E1 ubiquitin-activating enzyme, 0.4 mg/mL in 50 m$M$ HEPES, pH 7.6 (AFFINITI Research).
9. E2 ubiquitin-conjugating enzyme, UbcH7 (AFFINITI Research).
10. 2X Laemmli SDS-PAGE sample buffer containing 2-mercaptoethanol (added fresh).
11. SDS-PAGE equipment.
12. Blocking buffer: PBS containing 0.1% BSA plus 0.1% Tween-20.
13. Anti-ubiquitin monoclonal antibody (Sigma).
14. Anti-EGFR monoclonal antibody (cat. no. E12020, Transduction Laboratories, Lexington, KY).
15. Secondary rabbit anti-mouse antibody conjugated to HRP (Sigma).
16. ECL reagent (Amersham Pharmacia Biotech, Buckinghamshire, HP).

## 2.7. Examining Transient ERK Phosphorylation

1. Human kidney epithelial 293T cells.
2. LipofectAMINE™ reagent (Life Technologies, Inc., Carlsbad, CA).
3. Human recombinant EGF (Upstate Biotechnology, Lake Placid, NY).
4. RIPA buffer: 50 m$M$ Tris-HCL (pH 7.4), 150 m$M$ NaCl, 0.25 m$M$ EDTA (pH 8.0), 1% Triton X-100, 1% sodium deoxycholate, 0.2% sodium fluoride, 0.1% orthovanadate. Add 1X protease inhibitor cocktail (Boehringer Mannheim) to a final concentration of: 1 m$M$ PMSF, 0.5 μ$M$ sodium fluoride, 5 μg/mL aprotinin, 5 μg/mL leupeptin, 1 μg/mL pepstatin A, and 0.1% sodium orthovanadate (phosphatase inhibitor, SIGMA Chemicals) added before use.
5. Bio-Rad Protein Assay (Bio-Rad, Hercules, CA).
6. EGFR polyclonal conjugated to agarose (AC beads) (500 μg IgG/0.25 mL agarose) (cat. no. 1005, Santa Cruz Biotech, Santa Cruz, CA).
7. 2X Laemmli SDS-PAGE sample buffer containing 2-mercaptoethanol (added freshly).
8. SDS-PAGE equipment.
9. Anti-HA and anti-FLAG monoclonal antibodies (Roche Molecular Biochemicals, Indianapolis, IN).
10. Anti-EGFR monoclonal antibody (cat. no. E12020, Transduction Laboratories, Lexington, KY).
11. Anti-phospho-ERK1/2 and anti-pan-ERK monoclonal antibodies (Transduction Laboratories).
12. Secondary rabbit anti-mouse conjugated to HRP (Sigma).
13. ECL reagent (Amersham Pharmacia Biotech, Buckinghamshire, HP).

## 2.8. Examining Cell Surface Membrane-Associated EGFR

1. COS-1 monkey kidney cells.
2. LipofectAMINE™ reagent (Life Technologies, Inc., Carlsbad, CA).
3. Human recombinant EGF (Upstate Biotechnology, Lake Placid, NY).
4. PBS plus protease inhibitor cocktail (Boehringer Mannheim) added to a final concentration of: 1 m$M$ PMSF, 0.5 μ$M$ sodium fluoride, 5 μg/mL aprotinin, 5 μg/mL leupeptin, 1 μg/mL pepstatin A, and 0.1% sodium orthovanadate (phosphatase inhibitor, Sigma) added before use.
5. TE: 10 m$M$ Tris-HCl (pH 7.5) containing 0.1 m$M$ EDTA.
6. Homogenizing buffer: 1 m$M$ EDTA, 5 m$M$ HEPES (pH 7.5), 50 m$M$ sucrose.
7. 2X Laemmli SDS-PAGE sample buffer containing 2-mercaptoethanol (added fresh).
8. SDS-PAGE equipment.
9. Anti-HA and anti-FLAG monoclonal antibodies (Roche Molecular Biochemicals, Indianapolis, IN).
10. Anti-EGFR monoclonal antibody (cat. no. E12020, Transduction Laboratories, Lexington, KY).
11. Anti-G$_{i\alpha3}$ (3 rabbit antiserum (Upstate Biotechnology, Lake Placid, NY).
12. Rabbit anti-mouse antibody conjugated to HRP; goat anti-rabbit antibody conjugated to HRP (Sigma).
13. ECL reagent (Amersham Pharmacia Biotech, Buckinghamshire, HP).

## 2.9. Examining Cell Morphology and Long-Term ERK Activation

1. Rat pheochromocytoma (PC12) cells.
2. Dulbecco's modified Eagle's medium (DMEM), 4500 mg/mL glucose, complete growth medium: supplemented with 10 m$M$ HEPES (pH 7.4), 2 m$M$ glutamine, 100 U/mL penicillin/streptomycin, 10% heat-inactivated horse serum (Sigma), 5% FBS.
3. Collagen-coated plates (Iwaki® Chiba, Japan).
4. Poly-D-lysine hydrobromide—dilute and use at 0.1 mg/mL (Sigma).
5. Mammalian green fluorescence protein (GFP) expression constructs.
6. GenePORTER™ transfection reagent (Gene Therapy Systems, INC., San Diego, CA).
7. Human recombinant EGF (Upstate Biotechnology, Lake Placid, NY).
8. Crystal mount reagent (BiØmeda Co., Foster City, CA).
9. Bio-Rad MRC-1024 confocal laser optics attached to a microscope (Zeiss, Oberkochen, Germany) interfaced with an argon/krypton laser; LaserSharp software (Bio-Rad).

## 3. Methods

### 3.1. Confocal Laser Scanning Microscopy and Immunofluorescence Analysis to Visualize the Subcellular Localization of EGFRs

The process of receptor endocytosis has many phases, and it has been of of interest to investigate at which stage EGFR endocytosis is inhibited by hSpry2 overexpression, e.g., on the cell surface or at different stages of incorporation into vesicles (*see* **Fig. 1**).

1. Culture COS-1 cells in DMEM complete growth medium (*see* **Note 1**).
2. Trypsinize and spin down cells at 500$g$, resuspend in medium, count cells with hemocytometer. Suspend cells at $10^5$ cells/mL, seed 2 mL/well of a six-well plate, each with a sterile cover slip. Incubate cells at 37°C, 5% $CO_2$ for 18–24 h until cells are 60–70% confluent (*see* **Note 2**).
3. Prepare transfection mixture (per transfection/well): dilute 3–6 µL (use 3 µL per µg DNA) of LipofectAMINE reagent in an Eppendorf tube containing 250 µL Opti-MEM® I reduced serum medium. In another tube, dilute 1–2 µg (up to 10 µL) of DNA in 250 µL OptiMEM. Combine diluted LipofectAMINE reagent and DNA, mix well. Incubate at room temperature (RT) for 20 min. The solution may appear cloudy, but this will not impede transfection (*see* **Note 3**).
4. Meanwhile, wash cells once with OptiMEM and add 500 µL of OptiMEM to each well.
5. Mix gently and overlay the 500-µL lipid–DNA complexes onto the cells. Incubate in an incubator at 37°C with 5% $CO_2$ supply for 4–6 h.
6. Aspirate transfection medium and wash cells twice with complete growth medium. Add 2 mL fresh growth medium and incubate at 37°C in a 5% $CO_2$ incubator (*see* **Note 4**).
7. Following 16–24 h of incubation, wash cells twice with PBS and add 2 mL serum-free medium to each well for cell quiescence. Incubate at 37°C for overnight.

**EGFR**          **overlay**

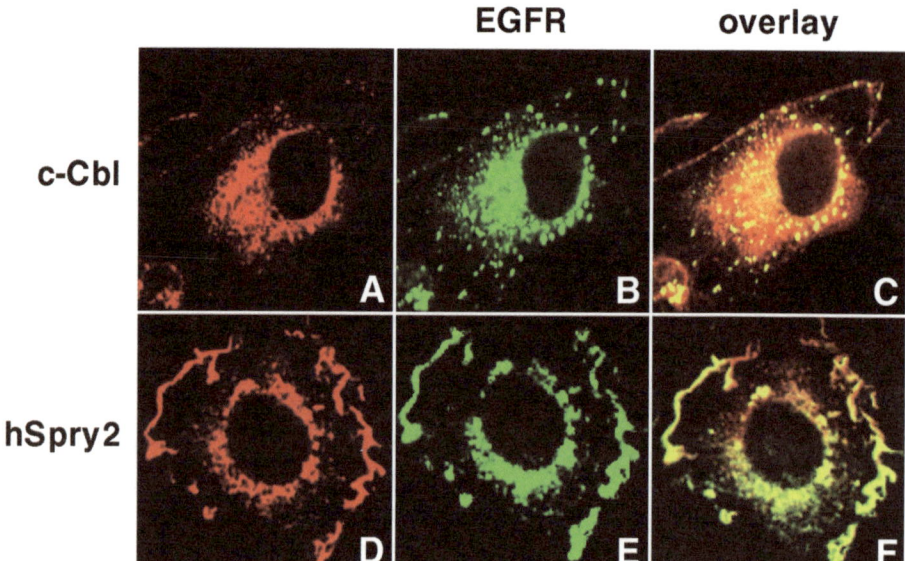

Fig. 1. Human Sprouty2 (hSpry2) inhibits epidermal growth factor receptor (EGFR) endocytosis, proteins are retained at the cell surface. COS-1 cells were singly transfected with 1 µg of HA-c-Cbl or FLAG-hSpry2, and subjected to serum-deprivation for overnight prior to stimulation with 100 ng/mL EGF at 37°C for 10 min. Cells were then fixed, permeabilized, and stained for endogenous EGFR/FITC. c-Cbl was visualized with anti-c-Cbl/Texas red, and FLAG-hSpry2 was detected using anti-FLAG/Texas red. Merged signals are labeled as overlay staining (yellow). In a representative EGF-stimulated cell, c-Cbl staining shows a punctate pattern indicative of its incorporation into endocytic vesicles (**A,C**); in the same cell, EGFRs can also be observed within those vesicles, with increased numbers on the cell periphery (**B,C**). Endocytic vesicular localization of EGFR can be counteridentified by staining with an antibody against EEA1, an established vesicle marker protein *(20)*. hSpry2 translocates to membrane ruffles upon EGF stimulation (**D**) and there appears to be a near total inhibition of EGFR endocytosis; receptors were retained at the membrane surface and not apparent in endocytic vesicles (**E,F**).

8. Leave quiescent cells untreated or stimulate with 100 ng/mL EGF at 37°C for 10 min, or other desired time period (*see* **Note 5**).
9. Rinse with PBSCM. Fix cells with 3% paraformaldehyde at 4°C for 30 min (*see* **Note 6**).
10. Wash twice with PBSCM, twice with PBSCM plus 100 m$M$ NH$_4$Cl, and twice with PBSCM.
11. Permeabilize cells with 0.1% saponin in PBSCM at RT for 15 min.
12. Primary antibody (1 µg) is diluted in 100 µL of FDB. Remove each cover slip, drain off excess liquid, and invert onto a clean plastic surface containing a 100-

µL droplet of the staining antibody. Incubate cover slip (cell layer facing down) at RT for 1 h.

13. Invert cover slip back into wells (cell layer facing up). Wash three times with PBSCM plus saponin.
14. Repeat **steps 12** and **13** for secondary antibody staining, but incubate in the dark.
15. Mount cover slip with crystal mount reagent and leave to dry at RT for overnight.
16. View slides: perform simultaneous double fluorescence acquisitions using 488 nm/568 nm laser lines to excite FITC/Texas red dyes; 40X oil immersion, 1.4 numerical aperture bright-field objective and fluorescein filter sets.

## 3.2. EGFR Downregulation Assay and Monenesin Inihibition of Receptor Recycling

Initial downregulation of EGFR was induced using an unlabeled ligand, and status of the remaining surface-associated binding sites was determined by performing a direct radiolabeled EGFR-binding assay (*see* **Fig. 2**). To circumvent the possibility that cell surface EGFR population is a result of the reappearance of recycled receptors, we include monensin, a chemical ionophore known to inhibit recycling of EGFRs *(14)*, in a repeated assay. Monensin acts to inhibit protein secretion.

1. Culture $10^5$ COS-1 cells/mL in 24-well plates (200 µL/well) in duplicates per transfection per time point. At 60–70% confluency, transfect cells with a total 1.5 µg of DNA, incubate for overnight before serum-starving the cells (*see* **Subheading 3.1., steps 1–7** and **Notes 1–4**).
2. Replace with fresh serum-free medium containing 50 µM of the proteasome inhibitor MG132 and 100 µM of chloroquine; pretreat at 37°C for 1 h (*see* **Note 7**).
3. Stimulate cells with unlabeled EGF (100 ng/mL) at 37°C in binding buffer (*see* **Notes 5** and **8**).
4. At the end of incubation, remove receptor-bound EGF remaining on the cell surface by a series of cold washes: three times with binding buffer, twice with ligand stripping buffer, and twice again with binding buffer. Add fresh binding buffer and let stand on ice for 5 min.
5. To determine the relative number of EGFR molecules (residual) on the cell surface, set up duplicate incubations with either 10 ng/mL $^{125}$I-EGF alone or $^{125}$I-EGF plus a 100-fold excess unlabeled EGF (to account for nonspecific binding); proceed at 4°C for 2 h.
6. Stop reaction by removal of unbound ligand. Rinse cell monolayers twice with cold binding buffer to remove nonsequestrated radioactivity.
7. Solubilize cells with 500 µL of 0.1 *M* NaOH containing 0.1% SDS at 37°C for 15 min (*see* **Note 9**).
8. Pipet resultant solutions into 2-mL screw-cap tubes and place in counter vials for analysis.
9. To quantify the receptor levels following inhibition of receptor recycling, preincubate cells at 4°C for 2 h with unlabeled EGF (100 ng/mL) in the absence

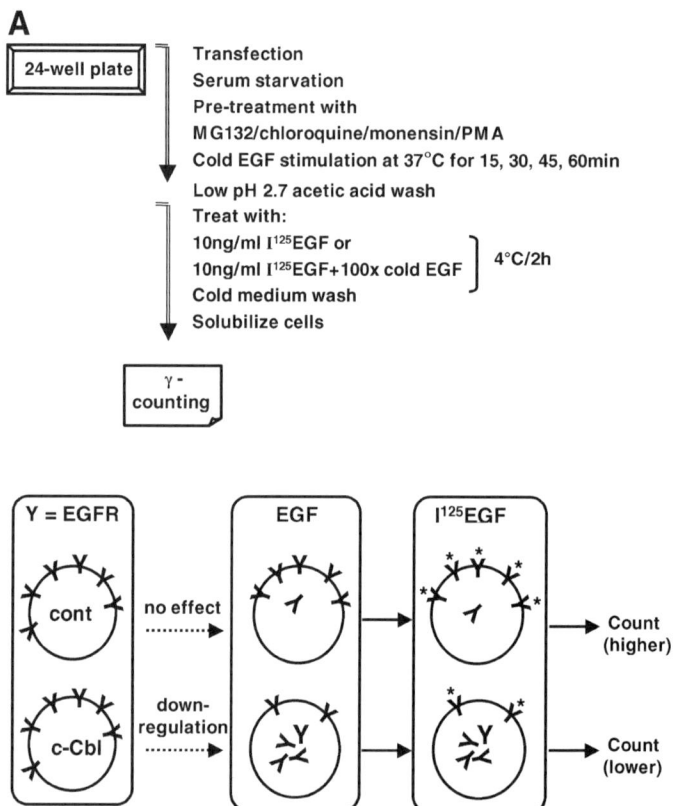

Fig. 2. Enhanced cell surface retention of EGFRs. (**A**) Summary of method to measure receptor downregulation.

Fig. 2. (**B**) *(opposite page)* To quantify surface EGFR population, COS-1 cells were cotransfected with 0.1 µg of an EGFR expression vector, together with 0.5 µg of plasmid encoding c-Cbl alone (▲) or with either 0.4 µg of FLAG-hSpry2 (◆) or FLAG-hSpry2ΔN11 (■) cDNAs, or with 1 µg plasmid encoding the RING Finger-defective form of c-Cbl (C381A) (▲). Forty-eight hours posttransfection, starved cells were stimulated without or with 100 ng/mL EGF at 37°C for various time periods as indicated. Bound EGF was removed and the level of surface EGFRs, relative to the initial number of ligand binding sites was determined by incubating sister cultures with $^{125}$I-EGF (10 ng/mL) at 4°C for 2–4 h, in the absence or presence of a 100-fold excess of unlabeled EGF (1 µg/mL). Control cells were not exposed to EGF (○). (**C**) COS-1 cells transiently expressing 0.2 µg EGFR were pre-treated with 100 n*M* PMA in the absence (△) or presence (▲) of 30 µ*M* monensin at 37°C. EGF (100 ng/mL) was added 20 min later. Following the indicated time periods, residual surface level of EGFR was determined in by using a radiolabeled ligand-binding assay. Control cells were not exposed to EGF or monensin (●).

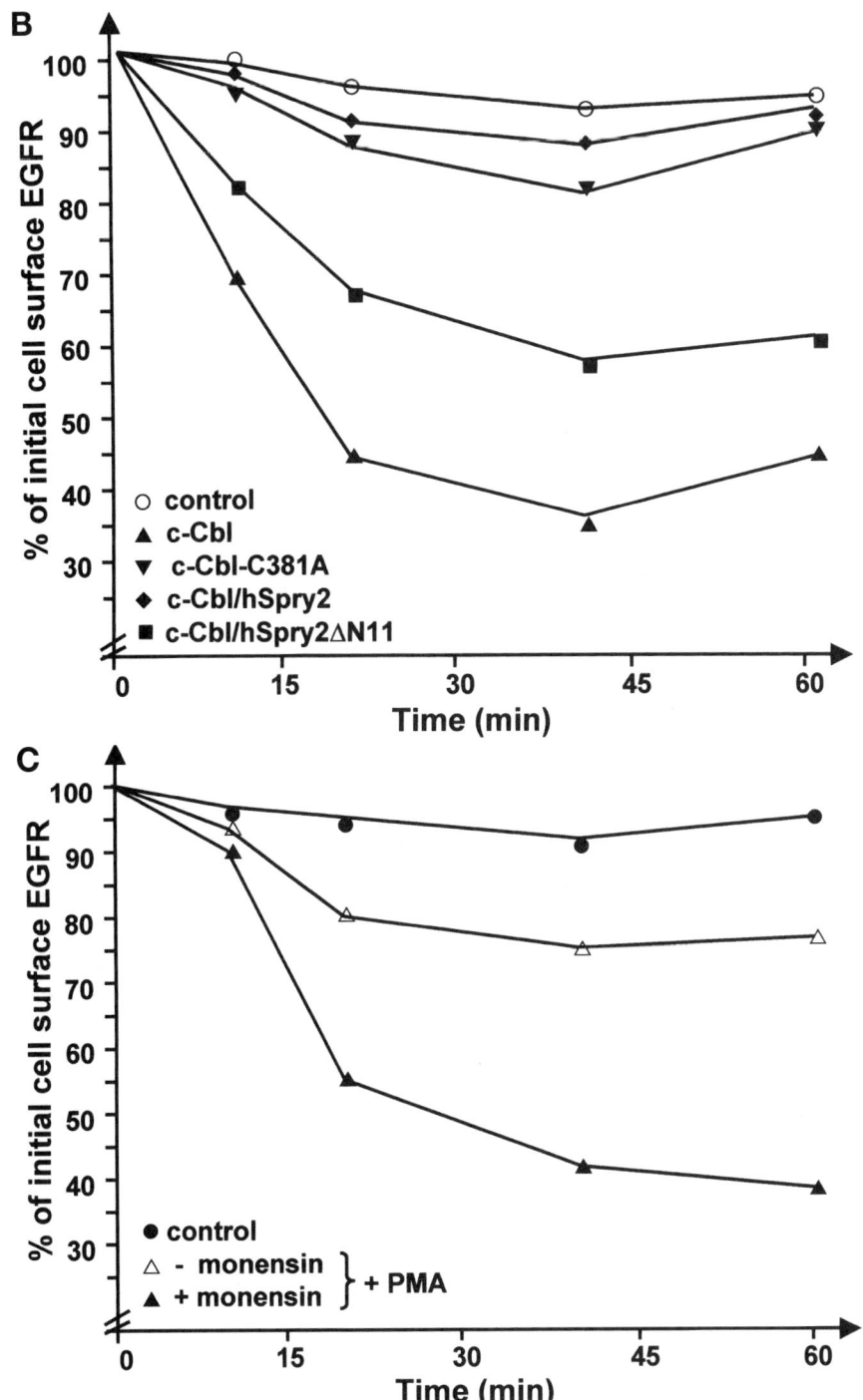

or presence of 30 μ*M* monensin. This incubation is followed by a temperature shift to 37°C for stimulation by EGF (**step 3**).

10. For induction of EGFR recycling, pretreat cells with 100 n*M* PMA for 20 min prior to EGF stimulation at 37°C. PMA has been established to divert internalized EGFR molecules from a degradative fate to a recycling pathway *(15)* (*see* **Note 10**).

11. After treatments in **step 9** or **10**, repeat the same procedure (**steps 3–8**).

### 3.3. Cell Surface Antigen Staining for Flow Cytometry to Measure Surface Retention of EGFR in Dual-Transfected Cells

This method measures cell-surface-retained EGFR (detectable by staining with antibodies against the extracellular epitopes of EGFR) and, specifically only for EGFR, hSpry2 dual-transfected cells. Nonpermeabilized cells were stained for both endogenous and overexpressed surface EGFR using red fluorescein labeling. Only cells that displayed dual fluorescence upon fluorescence-activated cell sorting (FACS) analysis were gated and assessed for levels of noninternalized, Texas-red-stained EGFR. The results indicate that hSpry2 inhibits EGFR endocytosis at a relatively early stage of the relocation/internalization process (*see* **Fig. 3**).

1. Seed cells in 35-mm dishes and grow them to 60–70% confluency before transfection (*see* **Subheading 3.1.**, **steps 3–7** and **Notes 1–4**) with respective amounts of EGFR and GFP expression constructs. Incubate overnight before serum starvation.

2. Rinse cells twice with PBS maintained in serum-free medium overnight prior to EGF treatment (100 ng/mL) at 37°C for the various time periods (*see* **Note 5**).

3. Lift cells off plating surface by treating with PBS/EDTA (*see* **Note 11**). Collect cells in an Eppendorf tube and spin down cells at 500*g* for 3 min at 4°C. Wash cells once in PBS.

4. Repeat centrifuge to pellet cells. Carefully remove supernatant (*see* **Note 12**).

5. Resuspend cell pellet at $2.5 \times 10^7$ cells/mL in cold assay buffer. Mix 100 μL of cells with 2 μg/100 μL diluted anti-EGFR antibody in an Eppendorf tube. Keep on ice for 45 min.

6. At the end of incubation, add 1 mL of assay buffer and spin-down cells at 500*g* (microfuge) at 4°C for 3 min. Remove supernatant, wash pellet three times with ice-cold assay buffer.

7. Resuspend cells in 100 μL ice-cold assay buffer, and mix with 100 μL of a 1/30 dilution of the Texas-red-conjugated secondary staining reagent (as in **step 5**). Cover samples with aluminium foil to shield from light in order to avoid bleaching of fluorophore. Repeat **step 6**.

8. Resuspend cells in 300 μL ice-cold PBS containing propidium iodide (1 μg/mL) for dead cell exclusion at 4°C for 10 min. Keep on ice and in the dark until ready for FACS analysis.

EGFR-TR

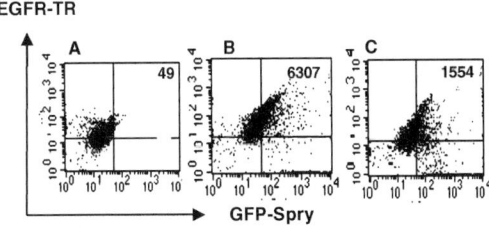

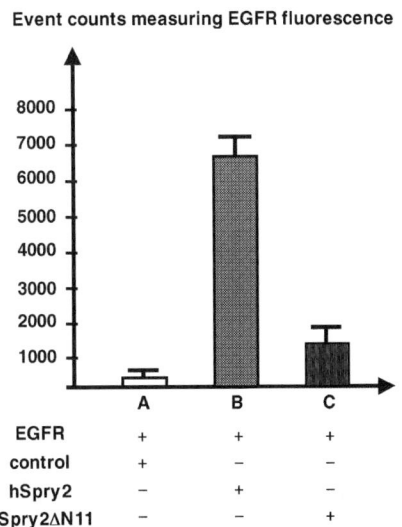

Fig. 3. Human Sprouty2 (hSpry2) abrogates c-Cbl-dependent ubiquitination of epidermal growth factor receptors (EGFRs). (**A**) COS-1 cells were co-transfected with 1 μg of HA-ubiquitin and 3 μg each of vector control, c-Cbl, c-Cbl-C381A, or the different FLAG-Sprys. Forty-eight hours later, cells were treated with 100 ng/mL EGF at 37°C for 10 min. Total cell lysates (TCL) were subjected to immunoprecipitation (IP) using anti-EGFR-AC beads, immunoblotted (IB) with anti-HA to distinguish ubiquitin-conjugated EGFR, and anti-EGFR to assess the amounts of IP EGFR. TCL blots were analyzed with anti-HA or anti-FLAG to show relative expression levels of the various constructs. The bracket indicates the position of high molecular species of ubiquitin-positive EGFRs. (**B**) IP EGFR proteins were subjected to an in vitro ubiquitination assay in the presence of purified UbcH7 (or no added UbcH7 as control without E2), eluted c-Cbl protein alone (or no added c-Cbl as control without E3), recombinant c-Cbl with either GST-hSpry2, GST-mSpry4, or GST-hSpry2ΔN11 fusion proteins, or c-Cbl-C381A fusion protein alone, plus the essential components in the ubiquitination system. The reaction products were analyzed via an immunoblotting protocol using anti-ubiquitin antibodies. The bracket highlights the position of high molecular species of ubiquitin-positive EGFRs.

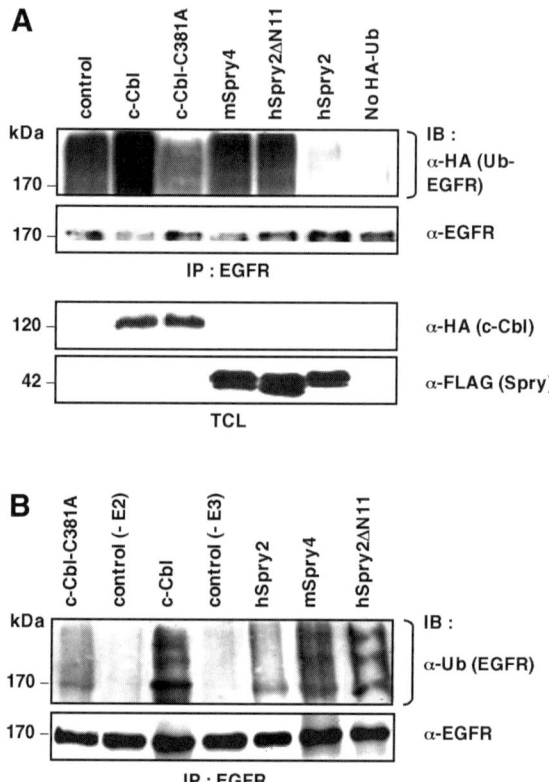

Fig. 4. Fluorescence-activated cell sorting (FACS) analysis to assess the surface epidermal growth factor receptor (EGFR) population in dual-transfected cells. GFP-hSpry2 or GFP-hSpry2ΔN11 were co-transfected into cells with EGFR. Cells were stained for surface EGFR using red-fluorescein labeling. Cells that displayed dual fluorescence were gated and assessed for levels of noninternalized EGFR. The number of GFP-hSpry2 (checked bar) co-expressing cells that contain high surface EGFR content was far greater than those co-expressing GFP-hSpry2ΔN11 (lined bar) or vector control.

9. Detection of plasmids encoding GFP: the GFP produced has an excitation peak at 470–490 nm and an emission peak at 510 nm. The expression level of GFP can be monitored by FACS analysis as described previously *(16)*.

### 3.4. Procedure for Detecting the Extent of In Vivo EGFR Ubiquitination

Expression of inactive RING Finger mutants of c-Cbl (C381A and ΔRF) inhibit EGFR ubiquitination and subsequent endocytosis *(4,17)*. As hSpry2 binds directly to the functional c-Cbl RING Finger necessary for the catalytic transfer of ubiquitin to substrate, this experiment addresses whether hSpry2 affects the ubiquitination status of EGFR *(see* **Fig. 4A***) (18)*.

1. Seed 293T cells in 100-mm dishes and grow for 24 h to 70–80% confluency, transfect cells with the respective amounts of each expression constructs and leave overnight before subsequent serum starvation (*see* **Subheading 3.1.**, **steps 3–7** and **Notes 1–4**).
2. Rinse cells with PBS and maintain in serum-free medium for 24 h prior to EGF treatment (100 ng/mL) at 37°C for 10 min (*see* **Note 5**).
3. After cell induction, discard growth medium. Wash cells with cold PBS three times. Discard PBS. Place the culture dish on ice. Lyse cells with RIPA buffer (0.5 mL/100-mm dish). Scrape lysed cells into Eppendorf tubes. Place on ice for 10 min.
4. Preclear lysates by microfuging at 16,000$g$ (microfuge) at 4°C for 10 min. Decant supernatant (total cell lysate) into new Eppendorf tubes. Aliquot 1–2 µL for Bio-Rad protein quantitation: 800 µL of purified $H_2O$ plus 200 µL of Bio-Rad solution plus 1–2 µL of total cell lysate (TCL), mix, read absorbance at 595 nm with a spectrophotometer. Extrapolate from a protein standard graph to determine the protein concentration (absorbance at 595 nm vs µg of BSA standard graph).
5. Immunoprecipitation: incubate equal amounts (~1 mg) of proteins (normalized from **step 5**) with 2.5 µg of agarose-conjugated anti-EGFR at 4°C for 4 h with constant rotation.
6. Quick-spin down IP beads at 500$g$ (microfuge). Wash beads three times using cold RIPA buffer.
7. Carefully decant supernatant. Add 2X Laemlli SDS-PAGE sample buffer and boil samples at 95°C for 5 min.
8. Resolve eluted proteins by SDS-PAGE (40 mA per gel for 1 h at RT). Electrophoretically transfer to polyvinylidene difluoride (PVDF) membranes (*see* **Note 13**).
9. Block membranes in PBS containing 1% BSA plus 0.1% Tween-20 for 1 h at 37°C, with shaking.
10. Immunoblot by incubating with 1 µg/mL primary antibody at RT for 1 h with shaking.
11. Wash membrane thrice with PBS containing 0.1% Triton-X-100, for 10 min each time.
12. As in **step 11**, incubate for staining with 1 µg/mL secondary antibody. Wash three times.
13. Detect immunoreactive protein bands using ECL.
14. To reuse membrane or to reprobe blot with other antibodies for normalization control: strip membrane in 0.8 mL of β-mercaptoethanol (added just before use) per 100 mL of stripping buffer at 60°C for 30 min, with gentle shaking. Rinse several times with purified $H_2O$ until the bubbles disappear. Repeat **steps 10–14** as necessary.

### 3.5. Procedure for Assaying EGFR Ubiquitination In Vitro

In this experiment, a ubiquitination reaction was reconstituted where EGFRs serve as substrates of c-Cbl in an in vitro enzyme mixture containing the essential ubiquitin system components. Three classes of enzymes are involved in the conjugation of ubiquitin to proteins. E1, the ubiquitin-activating enzyme, activates ubiquitin through an ATP-dependent formation of a high-energy thiol-ester bond between the C-terminal glycine carboxyl group of ubiquitin and the active-site cysteine within E1. This E1-activated ubiquitin is then transferred to a cysteine residue of an E2, or ubiquitin conjugating enzyme (UbC). An E2 enzyme, in conjunction with an E3 ubiquitin ligase, then transfers ubiquitin to a lysine residue on target protein, forming isopeptide bonds that constitute ubiquitin chain formation. Diverse combinations of E2-E3 complexes define substrate specificity. (For these studies, please refer to **Fig. 4B**.)

1. Elution of GST-recombinant proteins (pGex4T1 expression vector constructs):
   a. Add 2 mL of 50% slurry of glutathione Sepharose 4B (equilibrated with PBS) per 5 mg of GST in PBS + 0.1% Triton X-100. Incubate with gentle agitation at RT for 30 min.
   b. Centrifuge suspension at 500$g$ for 5 min to sediment the matrix. Remove supernatant.
   c. Wash pellet with 10 bed volumes of PBS (1 bed volume = 0.5 × volume of 50% GS4B slurry).
   d. Centrifuge suspension at 500$g$ for 5 min to sediment the matrix. Discard the wash.
   e. Repeat washing for a total of three washes.
   f. To the sedimented matrix, add 1 mL of elution buffer per each mL of bed volume of the glutathione Sepharose 4B slurry. Mix to resuspend matrix.
   g. Incubate at RT (22–25°C) for 10 min to elute the bound material from matrix.
   h. Centrifuge at 500$g$ for 5 min to sediment matrix. Remove supernatant into a new tube.
   i. Repeat elution and centrifugation steps twice more. Pool the three eluates.
   j. Determine yield of GST fusion protein. For GST, 1 $A_{280}$ = 0.5 mg/mL.
2. Endogenous EGFR was immunoprecipitated from COS-1 cells using EGFR-AC beads.
3. Incubate 2-μg aliquots of the protein-bound beads from **step 2** with 50 μL of reaction mixture containing the following: 1X reaction buffer, 10 μg ubiquitin, 100 ng E1 (*see* **Note 14**), 200 ng E2 (UbcH7), 1 μg E3 (GST-c-Cbl from **step 1**) or 1 μg c-Cbl-C381A (ubiquitination-defective mutant) and 1 μg of each eluted GST-tagged Spry fusion proteins (from **step 1**). Incubate enzymatic reactions at 37°C for 2 h.
4. Quick-spin at 500$g$ (microfuge) to pellet treated EGFR beads, wash three times using 1X reaction buffer.
5. Sample beads were boiled in 20 μL of 2X Laemmli SDS-PAGE sample buffer. Resolve reaction products on a high percentage (7.5%) SDS-PAGE gel. Transfer to membrane (*see* **Note 13**).
6. Block membrane in blocking buffer at RT for 1 h.
7. Western blot by incubating with 1 μg/mL anti-ubiquitin at RT for 1 h with gentle rotation.

8. Wash membrane thrice with PBS containing 0.1% Triton-X-100, 10 min each wash, with rotation.
9. As in **step 7**, stain with 1 µg/mL secondary antibody. Repeat washes as **step 8**.
10. Detect immunoreactive protein bands using the ECL.

### 3.6. Measurement of Short-Term ERK Phosphorylation Induced by EGFR Signaling

The pattern of MAPK activation can be directly visualized by immunoblotting using an antiserum specific for the doubly-phosphorylated form of ERK1/2 (pERK1/2). In addition, pERK1/2 staining directly reveals some features of signaling, i.e., the intensity of signal and the kinetics of RTK activation, which would be difficult to measure by other methods. The ability to visualize the output of RTK signaling also permits detailed establishment of epistatic relationships between signaling components of RTK cascades. There has been considerable debate whether receptor internalization occurs to propagate or terminate ERK signaling. (For these studies, please refer to **Fig. 5A**. These methods are similar to those under **Subheading 3.4.**, **steps 1–15**.)

### 3.7. Membrane Fractionation to Isolate Cell Surface Membrane-Associated EGFR

To correlate the surface receptor levels with ubiquitination, a time-course experiment was performed in which cell membrane fractions were extracted and the surface EGFR levels were compared. From the EGFR immunoblot shown in **Fig. 5B**, c-Cbl transfected cells show enhanced EGFR disappearance (with correspondingly lower amounts of surface EGFRs) over time compared to vector control cells; hSpry2-transfectants were shown to retain comparatively high receptor levels on the membrane across the same period of stimulation, in line with its blockade on ubiquitination of EGFRs. To serve as a loading control, a parallel blot was probed with an antibody specific for the heterotrimeric G protein $G_{i\alpha3}$ subunit, a membrane-bound protein shown to be unaffected by EGF treatment.

1. Harvest 60-mm dishes 293T cells by washing with 800 µL of PBS using a Pasteur pipet.
2. Spin cells out of medium at 500$g$ (microfuge) at 4°C for 3 min.
3. Reuspend cells in 200 µL of ice-cold PBS and collect cells as in **step 2**. Repeat one wash.
4. Freeze cell pellets at –80°C for 1 h (freeze-thaw step for easy lysis of cells).
5. Thaw out cell pellets on ice. Resuspend in 200 µL of homogenizing buffer.
6. Dounce homogenize cells with approx 25 strokes (for 3–4 min) slowly with twisting.
7. Spin cells at 500$g$ (microfuge) for 5 min at 4°C to remove nuclei and cell debris.
8. Transfer supernatant to fresh tubes and centrifuge at 16,000$g$ (microfuge) for 30 min at 4°C.

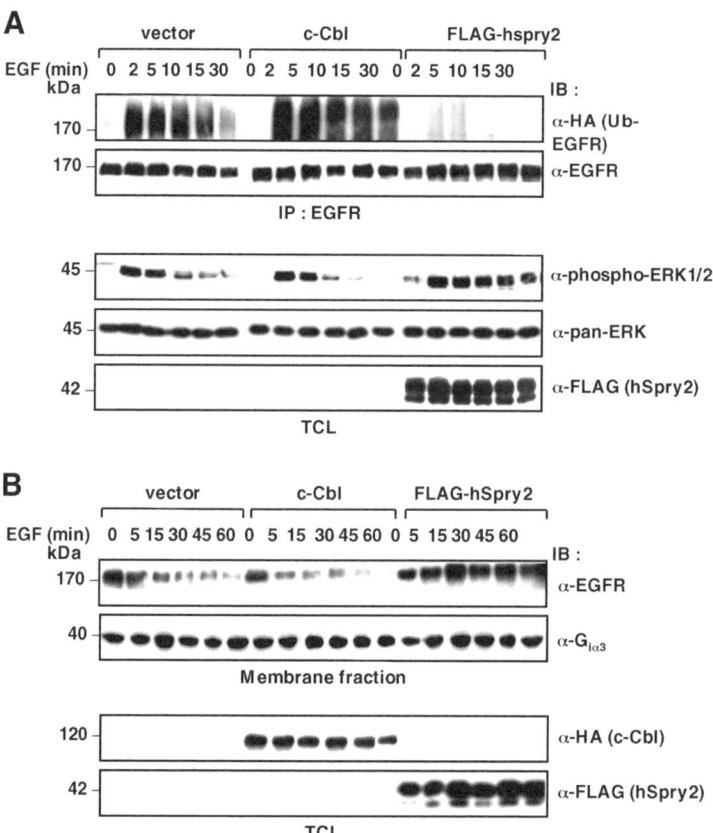

Fig. 5. Sustained ERK activity in response to prolonged epidermal growth factor (EGF) signaling at cell surface. **(A)** 293T cells were co-transfected with plasmids encoding EGF receptor (EGFR) (0.2 µg) and HA-ubiquitin (0.2 µg), together with a vector control, HA-c-Cbl (1.0 µg) or FLAG-hSpry2 (0.8 µg). Forty-eight hours posttransfection, cells were incubated with 100 ng/mL EGF at 37°C for the indicated time periods. Total cell lysates (TCL) were subjected to precipitation using anti-EGFR-AC beads, and receptor ubiquitination was detected with an anti-HA probe against HA-Ub-conjugated EGFR. An anti-EGFR blot shows equal amounts of IP EGFR. TCL blots were immunoblotted (IB) with anti-phospho-ERK1/2 (p42/44) to detect corresponding activated endogenous ERK levels, anti-pan-ERK to ascertain equivalent protein loading, and anti-FLAG to assess equality of expression of FLAG-hSpry2. **(B)** Cell membrane extracts of a similar set of transfections as depicted in panel **A** were isolated, and immunoblots of 1/10 of the sample volumes loaded were assayed with anti-EGFR to assess the levels of surface EGFR and with anti-$G_{i\alpha3}$ to ascertain equivalent amounts of loaded membrane fraction samples. TCL were analyzed for protein expression of transfected c-Cbl and hSpry2 by immunoblotting with anti-HA and anti-FLAG.

9. Decant supernatant into a new Eppendorf tube. Resuspend the final pellet containing the membrane fraction in 100 μL of TE buffer.
10. Use 5 μL of the membrane fraction to determine protein concentration by the Bio-Rad protein assay.
11. Add an equal volume of 2X Laemlli SDS-PAGE sample buffer to one-tenth of each fraction sample.
12. Resolve samples on 7.5% SDS-PAGE.
13. Perform immunoblotting similar to the procedures outlined under **Subheading 3.4., steps 10–15**.

## 3.8. Methods for Ascertaining That Sustained EGF-Induced ERK Activation Leads to Neurite Outgrowth in PC12 Cells

A good model system to examine MAPK responses in determining cellular phenotype is the proliferation/differentiation-responsive rat pheochromocytoma (PC12) cell line. There is well-supported evidence that the amplitude and longevity of MAPK signal governs whether these cells are stimulated to proliferate, or to withdraw from the cell cycle and differentiate; prolonged activation of ERK pathways (by FGF or NGF) in cultured PC12 cells would lead to their differentiation, as opposed to transient activation of the ERKs (EGF-induced) that leads to proliferation *(2,19)*. The PC12 cell system is enlisted here to demonstrate whether the sustained MAPK signal induced by EGF upon hSpry2 overexpression has biological consequences on cell fate. For this purpose, GFP-tagged Spry proteins are ideal for observing protein disposition in living cells. For these studies, please refer to **Fig. 6**.

1. PC12 cells grown on collagen-coated dishes (*see* **Notes 1, 2, 15,** and **16**). Coating of cover slips:
   a. Add 50 mL of sterile tissue culture grade water to 5 mg of polylysine.
   b. Coat cover slip with 1 mLof solution per 25 cm² surface area. Rock gently to ensure even surface coating.
   c. After 10 min, aspirate solution and rinse thoroughly three times with sterile tissue-culture-grade water.
   d. Allow cover slip to dry for at least 2 h before introducing cells and medium.
2. Trypsinize and spin down cells, resuspend in medium and count cells with hemocytometer. Suspend cells at $10^6$ cells/mL medium, plate in six-well plates (2 mL per well) containing poly-D-lysine-coated cover slips. Culture cells to 80–90% confluency before transfection.
3. Prepare transfection mixture: per transfection, dilute 10 μL (per μg DNA) of GenePORTER reagent in an Eppendorf tube containing 250 μL of serum-free medium. To a separate tube, dilute 1 μg of DNA into 250 μL of serum-free medium. Combine diluted transfection reagent and DNA. Mix gently and incubate at RT for 45 min.

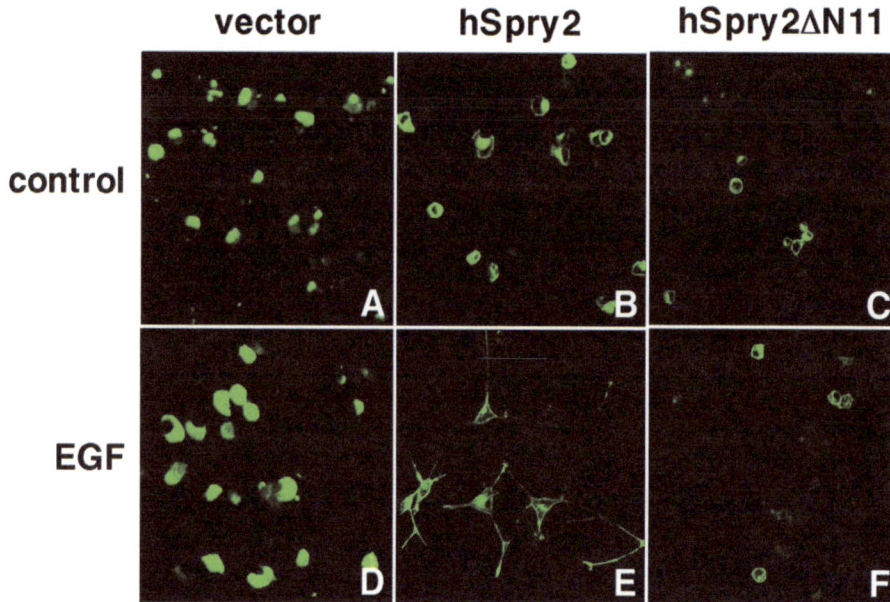

Fig. 6. Potentiation of epidermal growth factor receptor (EGFR) signaling leads to neuronal extensions. PC12 cells transfected with 1 μg of a plasmid encoding either GFP-vector control, GFP-hSpry2 or GFP-hSpry2ΔN11 were grown on polylysine-coated cover slips and fixed after 4 d of incubation in low-serum media without added growth factor (control; **A,B,C**) or supplemented with 100 ng/mL EGF (**D,E,F**). Panels show GFP-fluorescence (green) of representative cells from each experimental treatment.

4. Meanwhile, wash cells once with serum-free medium and add 500 μL of medium to each well.
5. Overlay the 500 μL of transfection mixture (from **step 3**) onto the cells and incubate at 37°C for 4–6 h.
6. Aspirate transfection medium and wash cells twice with complete growth medium. Add 2 mL of fresh growth medium with 10% FBS and incubate overnight at 37°C.
7. Following 16–24 h of incubation, wash cells twice with PBS and add 2 mL of serum-free medium to each well for cell quiescence. Incubate overnight at 37°C.
8. Leave starved cells untreated, or supplement medium with 100 ng/mL EGF. Incubate at 37°C for 4 d (*see* **Note 17**).
9. Proceed as under **Subheading 3.1.**, **steps 9–16**.

## 4. Notes

1. Selection of cell type for study: COS-1 cells are chosen for morphological studies due to their large and well-defined cellular structures; 293T cells are used in overexpression studies due to ease of handling and susceptibility to transfections;

CHO cells have no endogenous EGFR; A431 cells express high EGFR; and PC12 cells are used for their differentiation phenotype.

2. Because transfection efficiency is sensitive to culture confluency, it is important to maintain a standard seeding protocol from experiment to experiment. Recommended cell density for transfection: COS-1 at 60–70%; 293T at 70–80%; PC12 at 80–90%.

3. Transfection parameters are adjusted: (a) when transfecting with larger amounts of DNA and transfection reagent; and (b) when using larger tissue culture plates (in this case, increase the amounts of all reagents in proportion to the cell-plating area).

4. For most cell lines, add an equal volume of complete growth medium (with double the amount of FBS) to the transfected cells after 4–6 h of incubation without removing the transfection mixture. However, if toxicity is a problem, remove the transfection mixture and replace with fresh normal growth medium.

5. EGF stimulation in serum-free medium is best achieved in a low-volume medium, i.e., enough medium to just cover the surface of the cells. This issue is especially true for COS-1 cells. Additionally, basal EGFR phosphorylation can be reduced by prolonged starvation time, depending on the cell type being investigated. PC12 cells should be 70–80% confluent, i.e., one should decrease the cell density for longer treatments to allow further growth in serum.

6. Use ice-cold methanol to fix cells at 4°C for 5 min instead of paraformaldehyde to preserve the integrity of cell membranes for immunofluorescence staining procedures.

7. The final concentration of DMSO (used to dissolve MG132) in culture should not exceed 0.1% of the media volume. Subsequent buffer washes should include 20 μ*M* of MG132. *clasto*-Lactacystin β-lactone is another potent, highly specific and cell-permeable inhibitor of the proteasome. Lysosomal inhibitors include (with decreasing potency) chloroquine, leupeptin, and methylamine (Sigma).

8. EGF stimulation can be performed at 4°C to slow down receptor endocytosis.

9. Alternatively, use ice-cold STOP solution: 5% trichoroacetic acid.

10. Monensin is soluble in DMSO (25 mg/mL) and ethanol (25 mg/mL). The dose of recycling inhibitor administered was shown to be effective, evident in **Fig. 2C**, where cells exposed to PMA in concert with monensin show significantly lower levels of surface EGFRs compared to cells without monensin treatment.

11. Trypsinization detaches cells by nonspecific cleavage of cell surface proteins. Because this procedure will affect ligand binding to EGFR, the use of trypsin is not recommended for cell membrane fractionation studies. The alternative use of PBS/EDTA solution to detach cells is dependent on the cell type being investigated. Prolonged incubation times (longer than 10 min) will increase the amount of dead cells.

12. The cell pellet is not visible at this stage, so remove the supernatant carefully.

13. Transfer to membrane thoroughly using a constant 110 V source for 1.5 h. Most ubiquitin species are high molecular weight proteins. Overexpose blot to see ubiquitin laddering.

14. E1 is a critical component for the initation step in ubiquitin–protein isopeptide bond formation. The typical E1 concentration for conjugation in vitro ranges 50–200 n*M*. E1 is stable to multiple freeze/thaw cycles and can be stored at –80 or –20°C (for short periods).

15. Collagen plating: dissolve 100 mg of collagen (rat tail type I) in 1 L of 30% ethanol plus 70% PBS. Stir at 50°C in a water bath until dissolved. Pour enough solution to cover the plate surface and incubate for at least 2 h, but preferably overnight. Wash twice with PBS and dry under ultraviolet light.

16. For passaging of PC12 cells, manually dissociate the cells from the plate by pipeting medium over the collagen plate surface. If most cells have a nicely rounded morphology, use an attached yellow tip to provide greater force to ease dissociation from plate. If there are numerous abnormal, flattened greyish cells, do not use yellow tip so as to avoid dissociating the abnormal (differentiated) cells together with the normal ones.

17. Inhibitors of EGFR tyrosine kinase activity can be used during EGF treatment to test for the specificity of the EGF signaling pathway; recommended inhibitors include AG1478 and PD168393.

## References

1. Marshall, C. J. (1998) Signal transduction. Taking the rap. *Nature* **392,** 553,554.
2. Marshall, C. J. (1995) Specificity of receptor tyrosine kinase signaling: transient versus sustained extracellular signal-regulated kinase activation. *Cell* **80,** 179–185.
3. Yarden, Y. and Sliwkowski, M. X. (2001) Untangling the ErbB signaling network. *Nat. Rev. Mol. Cell Biol.* **2,** 127–137.
4. Levkowitz, G., Waterman, H., Zamir, E., et al. (1998) c-Cbl/Sli-1 regulates endocytic sorting and ubiquitination of the epidermal growth factor receptor. *Genes Dev.* **12,** 3663–3374.
5. Laney, J. D. and Hochstrasser, M. (1999) Substrate targeting in the ubiquitin system. *Cell* **97,** 427–430.
6. Hicke, L. (2001) Protein regulation by mono-ubiquitin. *Nat. Rev. Mol. Cell Biol.* **2,** 195–201.
7. Joazeiro, C. A. P., Wing, S. S., Huang, H. K., Leverson, J. D., Hunter, T., and Liu, Y. C. (1999) The tyrosine kinase negative regulator c-Cbl as a RING-type, E2-dependent ubiquitin-protein ligase. *Science* **286,** 309–312.
8. Casci, T., Vinos, J., and Freeman, M. (1999) Sprouty, an intracellular inhibitor of Ras signaling. *Cell* **96,** 655–665.
9. Tefft, J. D., Lee, M., Smith, S., et al. (1999) Conserved function of mSpry-2, a murine homolog of *Drosophila* Sprouty, which negatively modulates respiratory organogenesis. *Curr. Biol.* **9,** 219–222.
10. de Maximy, A. A., Nakatake, Y., Moncada, S., Itoh, N., Thiery, J. P., and Bellusci, S. (1999) Cloning and expression pattern of a mouse homologue of *Drosophila* Sprouty in the mouse embryo. *Mech. Dev.* **81,** 213–216.
11. Wong, E. S. M., Lim, J., Low, B. C., Chen, Q., and Guy, G. R. (2001) Evidence for direct interaction between Sprouty and Cbl. *J. Biol. Chem.* **276,** 5866–5875.

12. Lim, J., Wong, E. S. M., Ong, S. H., Yusoff, P., Low, B. C., and Guy, G. R. (2000) Sprouty proteins are targeted to membrane ruffles upon growth factor receptor tyrosine kinase activation: identification of a novel translocation domain. *J. Biol. Chem.* **275,** 32,837–32,845.

13. Wong, E. S. M., Fong, C. W., Lim, J., et al. (2002) Sprouty2 attenuates epidermal growth factor receptor ubiquitylation and endocytosis, and consequently enhances Ras/ERK signalling. *EMBO J.* **21,** 4796–4808.

14. Gladhaug, I. P. and Christoffersen, T. (1988) Rapid constitutive internalization and externalization of epidermal growth factor receptors in isolated rat hepatocytes. Monensin inhibits receptor externalization and reduces the capacity for continued endocytosis of epidermal growth factor. *J. Biol. Chem.* **263,** 12,199–12,203.

15. Bao, J., Alroy, I., Waterman, H., Schejter, E. D., Brodie, C., Gruenberg, J., and Yarden, Y. (2000) Threonine phosphorylation diverts internalized epidermal growth factor receptors from a degradative pathway to the recycling endosome. *J. Biol. Chem.* **275,** 26,178–26,186.

16. Cheng, L., Fu, J., Tsukamoto, A., and Hawley, R. G. (1996) Use of green fluorescent protein variants to monitor gene transfer and expression in mammalian cells. *Nature Biotechn.* **14,** 606–609.

17. Waterman, H., Levkowitz, G., Alroy, I., and Yarden, Y. (1999) The RING Finger of c-Cbl mediates desensitisation of the epidermal growth factor receptor. *J. Biol. Chem.* **274,** 22,151–22,154.

18. Mimnaugh, E. G., Bonvini, P., and Neckers, L. (1999) The measurement of ubiquitin and ubiquitinated proteins. *Electrophoresis* **20,** 418–428.

19. Kao, S. C., Jaiswal, R. K., Kolch, W., and Landreth, G. E. (2001) Identification of the mechanisms regulating the differential activation of the MAPK cascade by epidermal growth factor and nerve growth factor in PC12 cells. *J. Biol. Chem.* **276,** 18,169–18,177.

20. Mu, F. T., Callaghan, J. M., et al. (1995) EEA1, an early endosome-associated protein. EEA1 is a conserved alpha-helical peripheral membrane protein flanked by cysteine "fingers" and contains a calmodulin-binding IQ motif. *J. Biol. Chem.* **270,** 13,503–13,511.

# 6

## Dissecting the Epidermal Growth Factor Receptor Signal Transactivation Pathway

### Oliver M. Fischer, Stefan Hart, and Axel Ullrich

### Summary

Interreceptor cross-talk has emerged as a general concept in cellular signaling cascades. Therein epidermal growth factor receptor (EGFR) signal transactivation represents the so far best investigated cross-talk mechanism comprising heterogeneous receptor families. In this signaling process G protein-coupled receptor (GPCR) stimulation induces phosphorylation of the EGFR, combining the broad diversity of GPCRs with the potent signaling capacities of this receptor tyrosine kinase. Early reports attributed this transactivation mechanism to solely intracellular pathways as no EGF-like ligands could be detected in conditioned media of GPCR agonist-stimulated cells. However, Prenzel and colleagues demonstrated the involvement of metalloproteinase-mediated release of EGF-like ligands as the predominant mechanism of EGFR signal transactivation, providing a point of convergence for different intracellular effector proteins. Since this discovery, numerous investigations revealed the broad relevance of metalloproteinase-mediated ligand-dependent EGFR signal transactivation for coupling GPCRs to various cellular signaling responses. Here we describe methods to investigate GPCR-stimulated EGFR signal transactivation allowing the identification of both the EGF-like ligands and the metalloproteinases involved.

**Key Words:** EGFR; Heparin-binding epidermal growth factor (HB-EGF); EGF-like ligand; a disintegrin and a metalloproteinase (ADAM); GPCR; fluorescent activated cell sorting (FACS); cell surface shedding; small interfering RNA (siRNA); EGFR signal transactivation.

## 1. Introduction

Signal transduction networks utilize cell surface receptor cross-communication to achieve both signal diversification and integration depending on the cellular context.

From: *Methods in Molecular Biology, vol. 327: Epidermal Growth Factor: Methods and Protocols*
Edited by: T. B. Patel and P. J. Bertics © Humana Press Inc., Totowa, NJ

The discovery of epidermal growth factor receptor (EGFR) signal transactivation as a crucial mechanism combining the diverse cell surface receptor family of G protein-coupled receptors (GPCRs) and the potent signaling capacities of the EGFR *(1)* has led to numerous reports demonstrating the relevance of this cross-talk pathway in various cellular systems as well as in pathophysiological signaling situations such as cardiac hypertrophy, *Helicobacter pylori*-induced pathophysiological processes, cystic fibrosis, and cancer (reviewed in **ref. 2**).

Different reports implicated intracellular signaling mediators including cytoplasmic kinases such as c-Src, focal adhesion kinase (FAK), or protein kinase C (PKC), calcium levels or protein tyrosine phosphatases, in EGFR signal transactivation *(3,4)*. These investigations pointed towards diverging mechanisms depending on the cellular system while lacking a general mechanistic concept. However, a report by Prenzel and colleagues demonstrated that the molecular mechanisms of EGFR signal transactivation involve processing of transmembrane growth factor precursors by metalloproteinases *(5)*. This mechanism has been found in a wide variety of cellular systems, underlining its broad mechanistic relevance in EGFR signal transactivation. Depending on this ligand-dependent mechanism, diverse cellular responses were shown to be induced by GPCR stimulation, including cell proliferation, migration, or antiapoptotic responses in both physiological and pathophysiological signaling contexts *(2)*.

Recent reports implicated metalloproteinases of the a disintegrin and a metalloproteinase (ADAM) family in EGFR signal transactivation, as these proteinases have been frequently linked to the regulation of EGF-like ligand availability (reviewed in **ref. 6**). The EGF-like ligands used in the transactivation process described so far comprise the heparin-binding ligands heparin-binding EGF (HB-EGF) and amphiregulin and transforming growth factor (TGF)-α *(5,7,8)*. Because different combinations of ADAM proteinase and EGF-like ligand have been reported, the use of the respective proteinase–ligand combination appears to depend on the cellular system being investigated *(2)*. This chapter will focus on different methods used to investigate metalloproteinase-mediated EGFR signal transactivation and identify the components involved in this process, namely the EGF-like ligand and the metalloproteinase. (For additional discussion on approaches related to this topic, *see* Chapter 7.)

## 2. Materials

All of the methods described in this section require cell culture equipment and expertise in mammalian cell culture, polyacrylamide gel electrophoresis (PAGE), and Western blot equipment.

## 2.1. GPCR Ligands and Inhibitors

1. Cos7 cells.
2. αEGFR antibody, polyclonal, e.g., UBI (suitable for immunoprecipitation/Western blot detection).
3. Protein A sepharose, Sigma, mix 1:1 with sepharose.
4. Cell lysis buffer: 50 m$M$ hydroxyethyl piperazine ethane sulfonate (HEPES) (pH 7.5), 150 m$M$ NaCl, 1 m$M$ EDTA, 1% TritonX-100, 10% glycerin, 10 m$M$ $Na_2P_4O_7$.
5. HNTG buffer: 20 m$M$ HEPES (pH 7.5), 150 m$M$ NaCl, 0.1% TritonX-100, 10% glycerin, 10 m$M$ $Na_2P_4O_7$.
6. AG1478, EGFR kinase specific inhibitor, Alexis Corp., dissolve in dimethylsulfoxide (DMSO) at stock concentration of 10 m$M$, use at final concentration of 250 n$M$.
7. TAPI, metalloproteinase inhibitor, Calbiochem, dissolve in DMSO at stock concentration of 25 m$M$, use at final concentration of 10 μM.
8. Epidermal growth factor (EGF), Sigma.
9. Lysophosphatidic acid (LPA), Sigma, dissolve in phosphate-buffered saline (PBS) containing 0.1% bovine serum albumin (BSA) as a carrier.
10. PBS: 13.7 m$M$ NaCl, 2.7 m$M$ KCl, 80.9 $Na_2HPO_4$, 1.5 m$M$ $KH_2PO_4$, pH 7.4.
11. Laemmli sodium dodecyl sulfate-polyacrylamine gel electrophoresis (SDS-PAGE) equipment *(9)*.

## 2.2. Flow Cytometric Analysis

1. Assay buffer: PBS, 5% fetal calf serum (FCS), 0.005% sodium-azide.
2. HB-EGF antibody (goat), RnD-Systems (*see* **Note 10**).
3. Fluorescein isothiocyanate (FITC)-conjugated second antibody anti-goat, Sigma.
4. PBS/EDTA solution: PBS containing 10 m$M$ EDTA.
5. Propidium iodide solution, Sigma, 2 mg/mL, dilute in PBS 1:1000.
6. Flow cytometer.

## 2.3. Analysis of Mature EGF-Like Ligands in the Cell Culture Medium

1. Cos7 cells.
2. Mammalian expression construct for proHB-EGF.
3. Trichloroacetic acid (TCA) solution, 50%.
4. Na-DOC solution, sodium-desoxycholate, 10 mg/mL.
5. Schägger-Jagow tricine SDS-PAGE system *(10)*.
6. Monoclonal αHB-EGF antibody, RnD-Systems.

## 2.4. siRNA Oligonucleotide Treatment

1. oligo small interfering RNA (siRNA) Duplexes (synthesized, purified, and annealed at Dharmacon [*see* **Note 13**]) were dissolved at concentrations of 20 μ$M$ in water under sterile conditions and stored at −20°C. The siRNA sequences raised against EGF-like ligands and ADAM proteinases are listed in **Table 1**. gl2

### Table 1
### Sequences of Oligo Small Interfering RNAs Directed Against Epidermal Growth Factor-Like and "A Disintegrin and A Metalloproteinase" Proteins

| | |
|---|---|
| Control gl2 | AACGUACGCGGAAUACUUCGAdTdT |
| Amphiregulin | AACCACAAAUACCUGGCUAUAdTdT |
| | AAAAAUCCAUGUAAUGCAGAAdTdT |
| HB-EGF | AAGUGAAGUUGGGCAUGACUAdTdT |
| | AAUACAAGGACUUCUGCAUCCdTdT |
| TGF-α | AAAACACUGUGAGUGGUGCCGdTdT |
| | AAGAAGCAGGCCAUCACCGCCdTdT |
| ADAM10 | AAUAUUAUUAUGUGCCCCGUGdTdT |
| | AAACUUGGCUCUCAAUAAACTdTdT |
| ADAM12 | AACCUCGCUGCAAAGAAUGUGdTdT |
| | AAGACCUUGAUACGACUGCUGdTdT |
| ADAM17 | AAAGUUUGCUUGGCACACCUUdTdT |
| | AAAGUAAGGCCCAGGAGUGUUdTdT |

The sequences were selected according to the guidelines in **refs. *11*** and ***12***.

HB-EGF, heparin-binding epidermal growth factor; TGF, transforming growth factor; ADAM, a disintegrin and a metalloproteinase.

(directed against luciferase) is used as a control siRNA. For selection of siRNA sequences refer to **refs. *11*** and ***12***.

2. SCC-9 cells.
3. Transfection reagent Oligofectamine™ was purchased from Invitrogen (*see* **Note 15**).
4. Sterile polystyrene tubes (15 × 120 mm) were purchased from Becton-Dickinson Labware.
5. Polyclonal αADAM17/Tace antibody for Western blot detection (Chemicon).

## *2.5. Inhibition of Endogenous ADAM-Function With Dominant Negative ADAM Expression Constructs*

1. Primer for ADAM 17 wt-HA-tagged:
   5'-OLIGO-AD17WT: ACA GAA TTC GCC ACC ATG AGG CAG TCT CTC CTA TTC
   3'-OLIGO-AD17WT: TGC TCT AGA TTA AGC GTA ATC TGG AAC ATC GTA TGG GTA TCC TCC GCA CTC TGT TTC TTT GCT GTC
   5'-OLIGO-AD17DN: ACA GAA TTC GCC ACC ATG AGG CAG TCT CTC CTA TTC CTG ACC AGC GTG GTT  CTT TTC GTG CTG GCG AGC AAT AAA GTT TGT GGG  AAC TCG (high-performance liquid chromatography [HPLC] purified).

2. Enzymes: *Pfu*-polymerase, *Eco*RI, *Xba*I, CIAP, Ligase.
3. Retroviral expression construct, e.g., pLXSN-vector (BD Biosciences) and retroviral packaging cell line, e.g., Phoenix A cells (*see* **Note 16**).
4. 2X HEPES-buffered saline (HBS): 400 m$M$ NaCl, 50 m$M$ HEPES (sodium salt), 1.5 m$M$ Na$_2$HPO$_4$, bring up to 500 mL (pH 7.0).
5. CaCl$_2$ solution, 1 $M$.
6. Chloroquine, 25 m$M$ stock solution in PBS.
7. Polybrene, 4 mg/mL stock solution in PBS.
8. Monoclonal α-HA antibody suitable for Western blot detection (Babco).

## 3. Methods

### 3.1. Analyzing EGF-Like Ligands Involved in EGFR Signal Transactivation Using Inhibitors

1. Seed cells of interest in six-well plates and grow them for 24 h to 60–70% confluence.
2. Serum-starve cells for 24 h (*see* **Notes 1** and **17**).
3. Preincubate cells with AG1478 (250 n$M$) or TAPI (10 µ$M$) for 20 min (*see* **Note 2**).
4. Stimulate cells with LPA (10 µ$M$) and EGF (2 ng/mL) for 3 min.
5. Lyse cells using cell lysis buffer on ice for 10 min.
6. Collect lysates in Eppendorf cups and preclear lysates by centrifuging at 11,000$g$ in a cold centrifuge.
7. Immunoprecipitate EGFR by adding protein A sepharose and αEGFR antibody.
8. Incubate for 4 h at 4°C under constant rotation.
9. Spin down sepharose beads at 2800$g$ and wash three times using HNTG buffer.
10. After final wash step carefully remove supernatant and add Laemmli sample buffer (*9*).
11. Boil samples at 95°C for 5 min and subject them to gel ectrophoresis by standard Laemmli SDS-PAGE (*9*).
12. Transfer proteins to nitrocellulose by blotting.
13. Perform Western blot against phosphotyrosine residues.
14. Strip membrane.
15. Perform reblot against EGFR to control equal loading (*see* **Fig. 1**).

### 3.2. Flow Cytometric Analysis of EGF-Like Ligand Precursors

1. Seed Cos7 cells in six-well plates at 180,000 cells per well and grow cells for 24 h.
2. Serum-starve cells for 24 h by changing cell culture medium to serum free medium.
3. Stimulate with LPA (10 µ$M$) for 10 min (*see* **Notes 3** and **9**).
4. Wash cells with PBS.
5. Detach cells by removing the supernatant and adding 500 µL of PBS/EDTA solution per well (*see* **Notes 4** and **5**).
6. Incubate cells for 5–10 min in cell culture incubator.
7. Detach remaining adhesive cells by gentle shaking.
8. Collect cells in Eppendorf cups on ice.

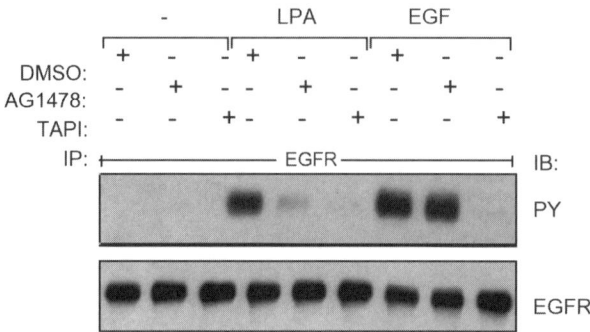

Fig. 1. Epidermal growth factor receptor (EGFR) signal transactivation in response to LPA stimulation. Cos7 cells were serum-starved for 24 h and preincubated with AG1478 (250 n*M*), TAPI (10 μ*M*), or an equal volume of dimethylsulfoxide (DMSO). Following stimulation for 3 min, cells were lysed and EGFR phosphorylation was assessed by immunoblot analysis.

9. Centrifuge at 500*g* at 4°C. The following steps are carried out maintaining the samples on ice.
10. Remove supernatant carefully (*see* **Note 6**) and resuspend cell pellet in 100 μL assay buffer containing primary αHB-EGF antibody (dilution: 2 μg/100 μL) by pipetting carefully up and down three to four times.
11. Incubate cell suspension with antibody for 45 min.
12. Prepare fresh Eppendorf cups containing 800 μL FCS.
13. Add 400 μL assay buffer to cell suspension and put the suspension carefully on top of the FCS.
14. Centrifuge cells at 500*g* for 5 min in a cold centrifuge (*see* **Note 7**).
15. Resuspend cells in 100 μL assay buffer containing FITC-conjugated secondary antibody (dilution: 1:1000) in the dark for 15 min.
16. Add 400 μL assay buffer to the cell suspension.
17. Prepare fresh Eppendorf cups with 800 μL FCS and put cell suspension carefully on top.
18. Centrifuge cells at 500*g* for 5 min in a cold centrifuge to remove unbound secondary antibody.
19. Resuspend cells in PBS containing propidium iodide to exclude dead cells from analysis.
20. Analyze cells on flow cytometer (*see* **Note 8**) (*see* **Fig. 2A**).

### 3.3. Analysis of Mature EGF-Like Ligands in the Cell Culture Medium

1. Seed Cos7 cells at 180,000 cells per well in a six-well plate.
2. The following day, transfect cells with an expression construct for proHB-EGF (*see* **Note 11**).
3. Change medium to serum-free cell culture medium and grow cells for another 24 h.

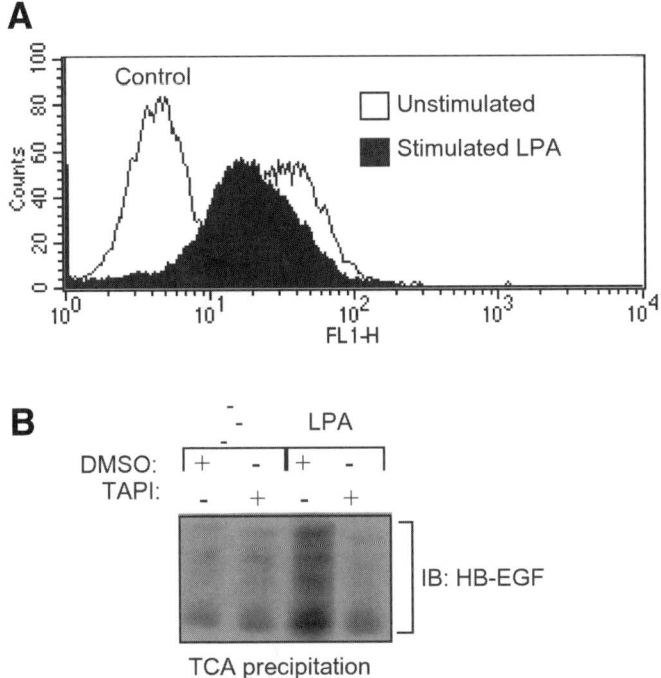

Fig. 2. Analysis of epidermal growth factor (EGF)-like ligand processing in response to EGF receptor (EGFR) signal transactivation. **(A)** Decrease of cell surface proHB-EGF content in response to EGFR signal transactivation. Cos7 cells were serum-starved for 24 h and subsequently stimulated with LPA for 10 min. Cells were treated as described in **Subheading 3.3.** and analyzed on a flow cytometer. **(B)** Release of mature heparin-binding EGF (HB-EGF) in cell culture supernatant. Cos7 cells were transfected with pcDNA3-HB-EGF-VSV and serum-starved for 24 h. After preincubation with TAPI or an equal volume of dimethylsulfoxide, cells were subsequently stimulated with lysophosphatidic acid (LPA) (10 μM) for 10 min. Conditioned media were treated as described under **Subheading 3.2.** HB-EGF content was analyzed by immunoblot analysis.

4. Exchange medium for 750 μL fresh cell culture medium per well (*see* **Notes 9** and **12**).
5. Stimulate with LPA (10 μM) and tetradecanoylphorbol acetate (TPA) (1 μM) as a positive control (*see* **Note 3**) for 10 min.
6. Collect cell culture supernatant in Eppendorf cups on ice.
7. Centrifuge at 11,000$g$ to remove any insoluble debris.
8. Add 7.5 μL of sodium-desoxycholate (10 mg/mL) and mix by inverting.
9. Add 250 μL of 50% TCA solution (final TCA concentration 10%) and mix by inverting.

10. Incubate on ice for 30 min.
11. Centrifuge supernatants at 11,000$g$ in a cold centrifuge at 4°C to separate precipitated proteins.
12. Discard supernatant carefully while not touching the precipitate.
13. Resuspend protein pellet in Schägger-Jagow sample buffer *(10)*.
14. Add Tris-HCl (pH 8.8) until the dye turns blue, indicating that the trichloroacetic acid is neutralized.
15. Separate samples on Schägger-Jagow tricine SDS-PAGE *(10)*.
16. Transfer proteins to nitrocellulose membrane.
17. Detect proteins by performing an immunoblot against HB-EGF (*see* **Fig. 2B**).

### 3.4. siRNA Oligonucleotide Treatment

The following protocol is designed for the transfection of cells in one well of a six-well tissue culture plate. For other applications use plate(s) of different size and adjust the reagents accordingly.

1. Trypsinize and spin down SCC-9 cells, resuspend in medium, and count the cell number. Suspend cells at $10^5$ cells/mL medium and plate in six-well plates (2 mL per well). Culture for 24 h before siRNA oligonucleotide treatment (*see* **Note 14**).
2. Prepare transfection solution A (for each transfection): dilute 2–4 μL of Oligofectamine™ reagent into medium without serum for a final volume of 15 μL. Allow diluted reagent to sit for 5–10 min.
3. Prepare transfection solution B (for each transfection): dilute 10 μL of a 20 μ$M$ stock siRNA oligonucleotide into 175 μL of medium without serum.
4. Add diluted Oligofectamine™ reagent (solution A) to diluted oligonucleotides (solution B), mix gently and incubate at room temperature for 15–20 min.
5. Wash cells once with medium without serum and add 800 μL of serum-free medium to each six-well plate.
6. Mix gently the solution of **step 6**, overlay the 190 μL of complexes onto the cells and incubate at 37°C for 4 h.
7. Add 1 mL growth medium containing two times the normal concentration of serum without removing the transfection mixture.
8. After 24-h incubation, wash cells once with PBS and add 2 mL of serum-free medium to each six-well plate and incubate for 24–48 h (*see* **Note 15**).
9. Stimulate cells with GCPR-ligands and EGF (1 ng/mL) for 3 min.
10. Assay lysates as described under **Subheading 3.1.**, **steps 5–15** (*see* **Fig. 3A,B**).

### 3.5. Inhibition of Endogenous ADAM-Function With Dominant Negative ADAM Expression Constructs

Overexpression of ADAM proteinases lacking the pro- and the metalloproteinase domain block the function of the endogenous ADAM proteinase and therefore act as dominant negative ADAM proteinases *(7)* (*see* **Fig. 4A,B**).

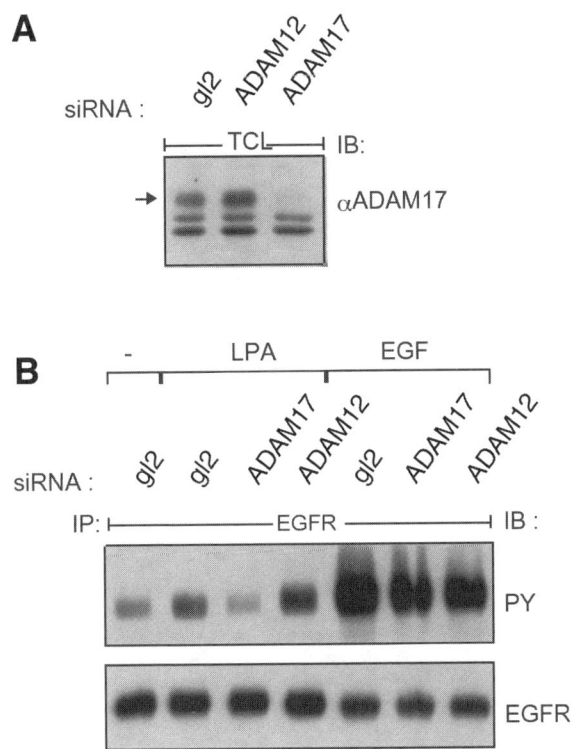

Fig. 3. **(A)** TACE small interfering RNA (siRNA) blocks endogenous ADAM17 expression. SCC-9 cells were transfected with ADAM12 and ADAM17 siRNA and protein level of TACE was analyzed by immunoblot with polyclonal anti-ADAM17 antibody. **(B)** ADAM17 siRNA inhibits epidermal growth factor (EGFR) signal transactivation in response to G protein-coupled receptor (GPCR) agonist treatment. SCC-9 cells were transfected with siRNAs raised against ADAM12 and ADAM17 and stimulated with agonists as indicated for 3 min. Following immunoprecipitation of cell lysates with anti-EGFR antibody, proteins were immunoblotted with antiphosphotyrosine antibody and reprobed with anti-EGFR antibody.

Construction of an eukaryotic expression plasmid of wild type (wt) and dominant negative (dn) ADAM17: wtADAM-17 cDNA is ligated into the retroviral expression vector pLXSN (BD Biosciences) (*see* **Note 16**) via the *Eco*RI and *Xba*I site after polymerase chain reaction (PCR) using placenta cDNA as template with a 5′ primer containing the starting ATG and a 3′ primer without stop codon but containing the codons for a HA-tag. The dominant negative ADAM17 construct is amplified by PCR using ADAM-17-wt as a template with a 5′ primer containing the signaling peptide and a 3′ primer without

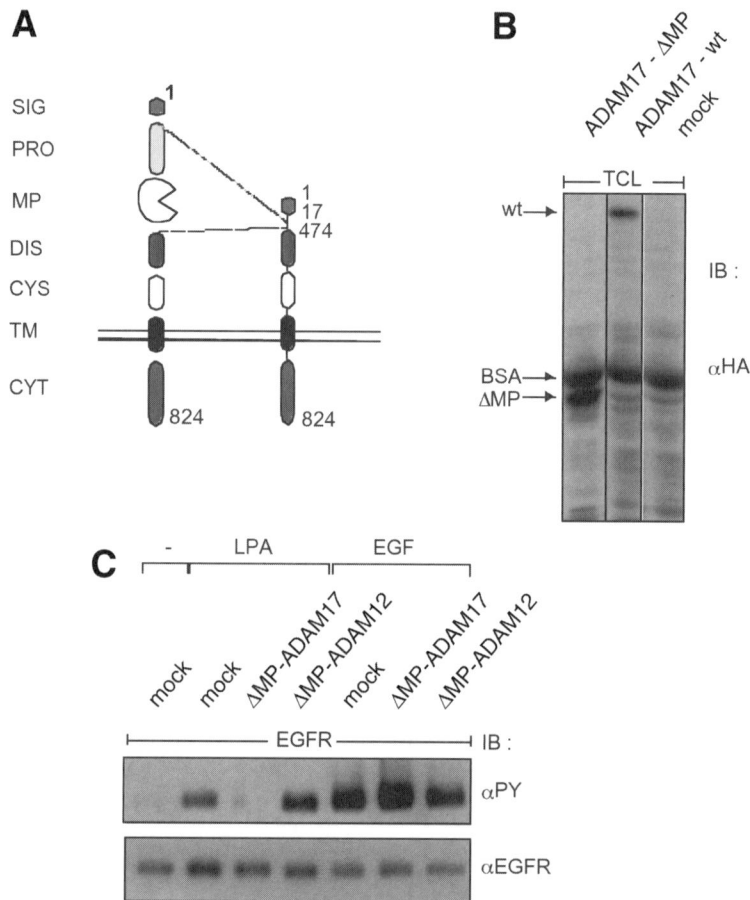

Fig. 4. **(A)** Schematic representation of human TACE. The structures of full-length human membrane-bound TACE and its dominant negative form are shown schematically. SIG, PRO, MP, DI, CYC, TM, and CYT represent the signal sequence, prodomain, metalloproteinase, disintegrin, cysteine-rich, transmembrane, and cytoplasmic regions, respectively. The amino acid positions of the full-length and the dominant negative form are shown on the right side of the construct. **(B)** Expression of wild-type and dominant-negative ADAM17 constructs. SCC-9 cells were infected with pLXSN vector (mock), ADAM17-HA wt or ADAM17-HA d.n. constructs. Forty-eight hours after infection cells were lysed and ADAM17 was detected by immunoblot analysis against the HA tag. **(C)** Dominant-negative ADAM17 blocks lysophosphatidic acid (LPA)-induced epidermal growth factor receptor (EGFR) signal transactivation in SCC-9 cells. SCC-9 cells were infected with dn constructs against ADAM12 and ADAM17 and stimulated with agonists as indicated for 3 min. Following immunoprecipitation (IP) of cell lysates with anti-EGFR antibody, proteins were immunoblotted (IB) with anti-phosphotyrosine antibody and reprobed with anti-EGFR antibody.

a stop codon but containing the codons for a HA-tag and ligated into the retroviral expression vector pLXSN (BD Biosciences) via the *Eco*RI and *Xba*I site. After transformation of an appropriate bacterial strain (DH5α, Invitrogen) and propagation of plasmid DNA in bacteria the correct insertion of the cDNA into the retroviral vector is confirmed by sequencing or restriction digestion.

### 3.5.1. Infection of SCC-9 Cells With Retroviral Expression Construct

1. Seed $2 \times 10^6$ Phoenix A cells in 60-mm plates (volume: 3 mL medium without antibiotics) and culture the cells for 24 h.
2. Add 5 min prior to transfection 2 µL chloroquin to each 60-mm plate.
3. For each plate, combine 8 µg of DNA, 370 µL of $H_2O$, 122 µL of 1 $M$ $CaCl_2$, and 500 µL of 2X HBS, vortex 10 s, and add HBS/DNA solution dropwise onto the 60-mm plate. Incubate cells for 9 h in S2-incubator.
4. Change medium and incubate till next morning.
5. Seed the target cells in low density in six-well plates.
6. Aspirate culture medium of the 60-mm plate and add 1.3 mL of fresh medium containing serum and incubate cells for 3 h.
7. Collect supernatant from Phoenix A cells, add 1.3 mL of fresh medium and incubate for 3 h.
8. Filter medium from **step 7** by 0.45 µm filter, add 1 µL polybrene per mL medium, and add onto target cells after aspiration of the old medium and incubate cells for 3 h.
9. Repeat **steps 7** and **8** two times.
10. Change medium of target cells and incubate for 24 h.
11. Aspirate medium of the target cells, wash cells with PBS and add 2 mL serum-free medium to each six-well plate.
12. Incubate target cells for 24 h (*see* **Note 17**) and analyze the cells for EGFR signal transactivation (*see* **Fig. 4C**).

## 4. Notes

1. Cells must be approx 70–80% confluent on the day of stimulation.
2. In addition to AG1478 and TAPI, the diphtheria toxin mutant CRM197 (Quadratech, UK, use at 5 µg/mL) is a useful tool to block specifically HB-EGF function. Heparin (Sigma, use at 100 µg/mL) interferes with both HB-EGF and amphiregulin function. The combined use of both inhibitors might therefore gain valuable insight into the EGFR signal transactivation mechanism.
3. TPA (12-*O*-tetradecanoylphorbol13-acetate, Sigma: dissolved in DMSO, stock concentration 1 m$M$, final concentration 1 µ$M$) induces unspecific cleavage of cell surface proteins, including EGF-like ligands, and can therefore be used as a positive control.
4. Trypsinization detaches cells by unspecific cleavage of cell surface proteins. As this procedure will also affect EGF-like ligands such as HB-EGF, we do not recommend the use of trypsin.

5. The use of PBS/EDTA solution to detach cells depends on the cell type being investigated. Because not all cells become detached within 10 min by this treatment, these cell types might not be used in this assay, as prolonged incubation times using PBS/EDTA longer than 10 min will increase the amount of dead cells.

6. The cell pellet is not visible, therefore, special care has to be taken when removing the supernatant.

7. Upon centrifugation cells will pass through the FCS layer. Thereby unbound antibody is removed and unspecific binding to the cells is reduced.

8. The instrument settings of the flow cytometer should be adjusted in such a way that the signal arising from the secondary antibody alone (background control) starts in the origin.

9. Inhibitors can be used to gain further insight into the mechanistic details of the signaling as described under **Subheading 3.1.** However, it is important to include the inhibitors in any change of media performed after preincubation of the cells with the respective inhibitory compounds.

10. The antibody used for flow cytometric detection of EGF-like ligand processing has to be tested carefully for effective cell surface staining since some antibodies give raise to a broad distribution of overall fluorescence intensity. As a consequence, a shift in fluorescence intensity towards lower fluorescence cannot be observed.

11. As the endogenous proEGF-like ligands are presumably retained in the extracellular matrix by binding to heparan-sulfate glycans, the pro-form of the ligand must be overexpressed to result in detectable amounts of mature ligand released into the cell culture medium.

12. Because basal EGF-like ligand shedding takes place, fresh cell culture medium has to be added, as otherwise the increase in the mature growth factor might be difficult to detect due to high background.

13. For reliable experiments we suggest purified and annealed duplex siRNA oligos, which are ready to use in transfection after addition of water.

14. Cells should be 30% confluent on the day of transfection.

15. Efficiency of target gene mRNA downregulation depends on different parameters: (1) As the half-lifes of different proteins are not the same, a time course after siRNA transfection, e.g., 24–96 h, is necessary to determine the time of optimal knockdown. (2) The efficiency of the mRNA knockdown of siRNAs varies and can be enhanced by mixing two or three siRNAs raised against the same target mRNA. (3) The transfection efficiency is often the most critical parameter and for each experiment the transfection efficiency has to be determined either by transfection of a established siRNA or by transfection of a CY2-labeled siRNA.

16. This protocol is designed for the use of the retroviral vector pLXSN (BD Biosciences) and the packaging cell line Phoenix A, but other retroviral systems can also be used for infection of the ADAM constructs.

17. Increased basal EGFR phosphorylation can be reduced by prolonged starvation time, depending on the cell type investigated.

## References

1. Daub, H., Weiss, F. U., Wallasch, C., and Ullrich, A. (1996) Role of transactivation of the EGF receptor in signalling by G-protein-coupled receptors. *Nature* **379,** 557–560.
2. Fischer, O. M., Hart, S., Gschwind, A., and Ullrich, A. (2003) EGFR signal transactivation in cancer cells. *Biochem. Soc. Trans.* **31,** 1203–1208.
3. Gschwind, A., Zwick, E., Prenzel, N., Leserer, M., and Ullrich, A. (2001) Cell communication networks: epidermal growth factor receptor transactivation as the paradigm for interreceptor signal transmission. *Oncogene* **20,** 1594–1600.
4. Hart, S., Gschwind, A., Roidl, A., and Ullrich, A. (2003) Epidermal Growth Factor Receptor Signal Transactivation. *NATO Sci. Series: II: Math. Phys. Chem.* **129,** 93–103.
5. Prenzel, N., Zwick, E., Daub, H., et al. (1999) EGF receptor transactivation by G-protein-coupled receptors requires metalloproteinase cleavage of pro HB-EGF. *Nature* **402,** 884–888.
6. White, J. M. (2003) ADAMs: modulators of cell–cell and cell–matrix interactions. *Curr. Opin. Cell. Biol.* **15,** 598–606.
7. Gschwind, A., Hart, S., Fischer, O. M., and Ullrich, A. (2003) TACE cleavage of proamphiregulin regulates GPCR-induced proliferation and motility of cancer cells. *EMBO J.* **22,** 2411–2421.
8. McCole, D. F., Keely, S. J., Coffey, R. J., and Barrett, K. E. (2002) Transactivation of the epidermal growth factor receptor in colonic epithelial cells by carbachol requires extracellular release of transforming growth factor-alpha. *J. Biol. Chem.* **277,** 42,603–42,612.
9. Laemmli, U. K. (1970) Cleavage of stuctural proteins during the assembly of the head of bacteriophage T4. *Nature* **227,** 680–685.
10. Schagger, H. and von Jagow, G. (1987) Tricine-sodium dodecyl sulfate-polyacrylamide gel electrophoresis for the separation of proteins in the range form 1 to 100 kDa. *Anal. Biochem.* **166,** 368–379.
11. Elbashir, S. M., Harborth, J., Lendeckel, W., Yalcin, A., Weber, K., and Tuschl, T. (2001) Duplexes of 21-nucleotide RNAs mediate RNA interference in cultured mammalian cells. *Nature* **411,** 494–498.
12. Tuschl, T., Elbashir, S. M., Harborth, J., and Weber, K. www.rockefeller.edu/labheads/tuschl/sima.html (revised May 6, 2004).

# 7

## A Sensitive Method to Monitor Ectodomain Shedding of Ligands of the Epidermal Growth Factor Receptor

Umut Sahin, Gisela Weskamp, Yufang Zheng, Valerie Chesneau, Keisuke Horiuchi, and Carl P. Blobel

### Summary

All ligands of the epidermal growth factor receptor (EGFR) are made as membrane anchored precursors that can be proteolytically processed and released from the plasma membrane. This process, which is referred to as protein ectodomain shedding, is emerging as a key regulator of the function of EGFR ligands. In light of the important roles of EGFR signaling in development and disease, it will be important to understand more about the regulation of proteolytic processing of EGFR ligands. This chapter describes a sensitive and semiquantitative method to measure ectodomain shedding of EGFR ligands that was designed to facilitate studies of this process in cells.

**Key Words:** Epidermal growth factor receptor (EGFR); EGFR ligands; ectodomain shedding; metalloproteinases; disintegrins; a disintegrin and a metalloproteinase (ADAM); alkaline phosphatase-based reporter system.

## 1. Introduction

The epidermal growth factor receptor (EGFR) is a tyrosine kinase with critical roles in cell survival, proliferation, and differentiation in development and disease *(1,2)*. The EGFR is activated via binding of a set of ligands to its extracellular domain, and subsequent dimerization of the receptor leads to phosphorylation and activation of its downstream signaling pathways *(1,3–5)*. Seven ligands of the EGFR have been identified to date: epidermal growth factor (EGF), transforming growth factor (TGF)-α, amphiregulin, epiregulin, heparin-binding EGF-like growth factor (HB-EGF), betacellulin, and epigen *(5)*. Each of these EGFR ligands is synthesized as a membrane anchored precursor that can be proteolytically released from cells to generate soluble, biologically

From: *Methods in Molecular Biology, vol. 327: Epidermal Growth Factor: Methods and Protocols*
Edited by: T. B. Patel and P. J. Bertics © Humana Press Inc., Totowa, NJ

active signaling molecules. Even though membrane-bound EGFR ligands can engage in juxtacrine signaling (cell–cell signaling) *(6,7)*, release from the membrane not only increases their signaling range, but it may also facilitate EGFR dimerization and signaling *(8)*. In fact, metalloproteinase-dependent release of EGFR ligands has been shown to be critical for activation of the EGFR under variety of circumstances. For example, EGFR-dependent cell migration and proliferation can be blocked with a metalloproteinase inhibitor *(9)*, and crosstalk between G protein-coupled receptors and the EGFR also depends on processing and release of EGFR ligands *(10,11)*. Furthermore, mice lacking the membrane-anchored a disintegrin and a metalloproteinase 17 (ADAM17) protein have a phenotype that resembles that of mice lacking the EGFR, or two of its ligands (TGF-α, HB-EGF) *(12,13)*. Taken together, these results suggest that release of soluble EGFR ligand extracellular domains (ectodomains) from their membrane-tethered precursors can be critical for EGFR activation, at least in the cases mentioned above.

In light of the key role of ectodomain shedding in activating TGF-α and HB-EGF, and because the EGFR signaling pathway is a validated target for cancer therapy *(2)*, it will be important to learn more about the regulation of enzymes responsible for shedding EGFR ligands. We have recently used cell-based assays to uncover distinct roles for ADAM10 and ADAM17 in ectodomain release of six EGFR ligands from mouse embryonic cells *(14)*. In this chapter we discuss the design of these experiments and the considerations that went into devising what we consider a sensitive and semiquantitative approach to evaluate shedding of all currently known EGFR ligands. In addition, the reader is referred to Chapter 6 for a further discussion of ligand processing and its linkage to the dissection of EGFR-mediated signal transactivation pathways.

Three aspects of the procedures described below were particularly important for identifying which ADAMs are critical for shedding ligands of the EGFR in mouse embryonic cells. The first was the use of an alkaline phosphatase-based reporter system, which provides a simple and sensitive means to monitor shedding of EGFR ligands in cell-based assays. The second was the use of primary mouse embryonic cells lacking one or more ADAMs that are considered to be potential sheddases for EGFR ligands in these cells. The third was the use of a single-well shedding assay, in which the increase or decrease in ectodomain shedding in one well following addition of an inhibitor or activator of ectodomain shedding is compared to shedding in the same well under constitutive conditions. These three approaches are briefly outlined below.

## 1.1. Alkaline Phosphatase Reporter System

One of the problems in studying shedding of mouse EGFR ligands is that few antibodies are available for their detection by Western blot analysis. In addition, EGFR ligands are potent signaling molecules and do not need to be released in high amounts to activate their receptor. Once released, the soluble ligand may be captured by the EGFR, further lowering its concentration in the cell supernatant. The combination of these factors makes it inherently difficult to detect shedding of overexpressed mouse EGFR ligands. In order to circumvent this problem, we used alkaline phosphatase (AP)-tagged versions of all EGFR ligands for shedding assays. The shedding levels can thus be assessed by spectrophotometric detection of AP activity in the culture supernatant or by running the released proteins on a gel and visualizing the renatured AP in the gel through substrate addition (**Fig. 1**) (*see also* **ref.** *15*).

Use of primary mouse embryonic cells for shedding assays: Another important objective of our study was to use a "loss-of-function" approach to evaluate the contribution of different ADAMs to the shedding of EGFR ligands in cell based assays. For this purpose, we isolated cells from knockout mice for various ADAMs that are considered to be potential EGFR ligand sheddases *(14)*. Because ADAMs 9, 10, 12, and 17 had already been implicated in EGFR ligand shedding (*see* **ref.** *14* for details), we considered any ADAM that contains a catalytic site consensus sequence (HEXXH) and is expressed in a variety of somatic tissues as a potential EGFR ligand sheddase (i.e., ADAMs 9, 10, 12, 15, 17, and 19). Initially, we used immortalized mouse embryonic fibroblasts from different ADAM knockout mice for shedding studies. However, the expression levels of ADAMs and other genes varied considerably among different clones of immortalized cells, even if they were derived from the same animal (G. Weskamp and C. Blobel, unpublished results). Therefore, we decided against using immortalized cells for comparative "loss of function" studies of EGFR ligand shedding. Instead, we chose to work with primary mouse embryonic cells (referred to as mouse embryonic fibroblasts [mEF]) (*see* **Note 1**) for shedding studies in all cases where the corresponding knockout mice survive long enough to allow production of fibroblasts from embryos at embryonic day (E)13.5. The only exception were cells lacking ADAM10, which had to be immortalized because *adam10–/–* mice die at E9.5, too early to generate sufficient numbers of primary embryonic cells for these studies *(16)*. It is important to note that Northern and Western blot analyses confirmed that all candidate EGFR ligand sheddases (ADAMs 9, 10, 12, 15, 17, and 19) are expressed in wild-type mEFs, and that the wild-type mRNA was missing in cells isolated from animals that were null for the respective ADAM locus *(14)*. Experiments involving primary mEF isolated under identical conditions from wildtype and knockout embryos are well suited to allow direct comparison of shedding levels in the presence or absence of any gene of interest.

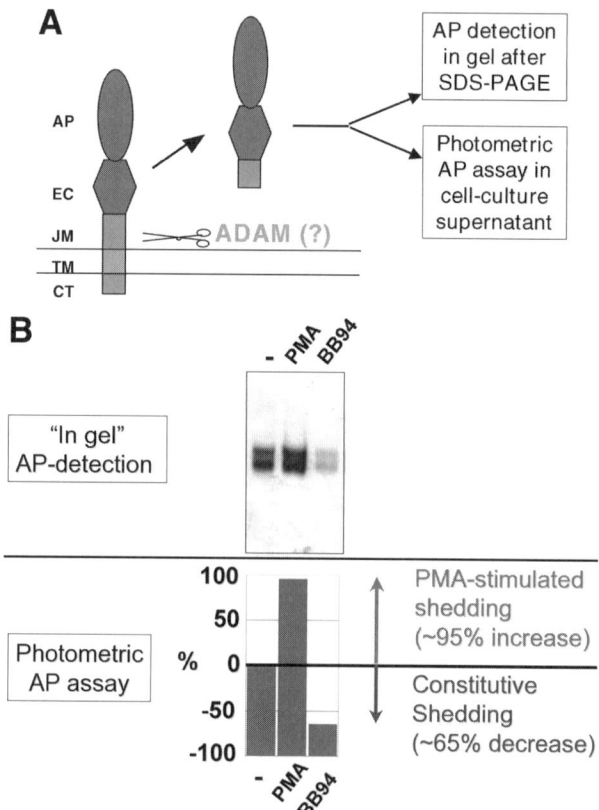

Fig. 1. An alkaline phosphatase detection system to monitor shedding of epidermal growth factor receptor (EGFR) ligands. A. Diagram of a fusion protein containing an alkaline phosphatase (AP) domain attached to the N-terminus of the membrane proximal EGF module of an EGFR ligand (after removal of the signal sequence). When an EGFR ligand–AP fusion protein is expressed in cells, shedding of its ectodomain by ADAMs (or other enzymes) into the culture supernatant can be detected in two ways: (1) The AP fusion protein can be enriched using the lectin Con A, run on an sodium dodecyl sulfate-acrylamide gel, and visualized following renaturation and color development, or (2) the AP level can be quantitated by spectrophotometric determination of AP activity in the culture supernatant (*see* text for details). Because the only AP activity in the cell supernatant derives from the shed fusion protein, there is little or no background staining in an in-gel AP assay B. Examples of an in-gel detection of transforming growth factor (TGF) shed from wild-type mouse embryonic fibroblasts and the corresponding spectrophotometric assay. The sample in the first lane corresponds to AP-TGF( released within 1 h from untreated cells (-), the sample in lane 2 was collected from the same well after 1 h of stimulation of shedding with the phorbol ester PMA, whereas the sample in lane 3 was collected from the same well after 1 h of treatment with the hydroxamic-acid-type metalloproteinase inhibitor batimastat. The experiment shown in this figure depicts three separate treatments of the same well and was prepared for illustration purposes only. Usually only two sequential treatments of cells in a given well are performed (as outlined in **Fig. 2**). Abbreviations: AP, alkaline phosphatase; EC, ectodomain (EGF module); JM, juxtamembrane domain; TM, transmembrane domain; CT, cytoplasmic domain; ADAM, a disintegrin and a metalloproteinase; PMA, phorbol-12-myristate-13-acetate; BB94, batimastat.

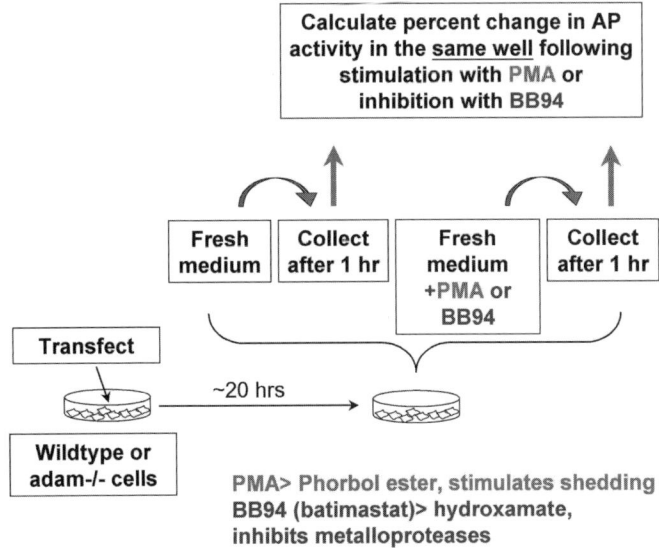

Fig. 2. Single-well shedding assay. The single-well shedding assay is a ratiometric assay that allows determination of the percent increase or decrease in ectodomain shedding of a given ligand following treatment with stimulatory or inhibitory compounds. A commonly used stimulator of ectodomain shedding is the phorbol ester phorbol-12-myristate-13-acetate (PMA), although any other stimulus, such as activators of G protein-coupled receptors, can be used. Proteinase inibitors are usually used to inhibit ectodomain shedding (such as the hydroxamic-acid-type metalloproteinase inhibitor batimastat [BB94]). The assay works as follows: tissue culture cells, such as primary mouse embryonic fibroblasts, are transfected with the alkaline phosphatase (AP)-tagged EGFR ligand. After a recovery period of 20 h, fresh medium is added and collected after 1 h. Then a second aliquot of fresh medium including the stimulatory or inhibitory reagent (such as PMA or BB94) is added, and collected after an incubation period of 1 h. The percent change in AP activity in the same well following treatment is calculated by comparing the AP activity released into the supernatant in the treated sample to the activity found in the supernatant of untreated cells.

## 1.2. Single-Well Shedding Assay

Despite the evident advantages of working with mEF cells, one concern was that it might be difficult to compare results from separate shedding experiments with cells from different knockout mice or different litters as a result of possible variations in transfection levels. Even though we ultimately found that the transfection levels are quite similar from experiment to experiment, we employed a single-well shedding assay for collecting all data points (**Fig. 2**). This means that two samples are collected sequentially from any given well of cells (wild-type or *adam*–/– mEF transfected with one of the EGFR ligands).

The first sample corresponds to culture supernatant that has been conditioned for 1 h to establish the levels of constitutive shedding from a certain well. Then fresh medium containing compounds that either stimulate ectodomain shedding (such as the phorbol ester phorbol-12-myristate-13-acetate [PMA], or the phosphatase inhibitor pervanadate) or inhibit it (such as the hydroxamate-based metalloproteinase inhibitor BB94) is added to the same well and conditioned for an additional hour. Each single-well assay thus provides one data point to indicate the percent stimulation or inhibition of shedding in the second sample compared to the baseline constitutive shedding in the first sample.

## 2. Materials

### 2.1. Preparation and Culturing of Primary Mouse Embryonic Fibroblasts

1. mEF growth medium: Dulbecco's modified Eagle's medium (DMEM) + 10% fetal calf serum (FCS) + penicillin/streptomycin.
2. Plain DMEM: DMEM; no FCS, no penicillin/streptomycin added.
3. PBS$^{-Ca-Mg}$: phosphate-buffered saline (PBS) without calcium/magnesium.
4. 0.25% trypsin/ethylenediaminetetraacetic acid (EDTA).
5. Opti-MEM (Gibco).
6. DNAseI (Roche).

### 2.2. Shedding Assays

1. PMA (Sigma).
2. Batimastat (BB94) (*see* **Note 9** for commercial sources of related hydroxamic acid-type metalloproteinase inhibitors).
3. Cell lysis buffer: PBS + 1% Triton X-100 + Proteinase inhibitor cocktail (Roche).
4. 2.5% Triton X-100.
5. Concanavalin A (ConA) (Sigma).
6. Elution buffer: 50 m$M$ Tris-HCl (pH 8.0), 0.5 $M$ α-D-methyl mannoside.
7. Nitroblue tetrazolium-5-bromo-4-chloro-3-indolylphosphate (NBT/BCIP) tablets (Roche) dissolved in 10 mL of water according to the manufacturer's instructions.
8. 4-Nitrophenyl phosphate (4-NPP, Roche; *see* **Note 15**).
9. Alkaline phosphatase buffer: 100 m$M$ Tris-HCl (pH 9.5), 100 m$M$ NaCl, 20 m$M$ MgCl$_2$.
10. 50% Methanol–10% glacial acetic acid.

## 3. Methods

Plasmids encoding AP-tagged EGFR ligands were constructed by S. Higashiyama by inserting partial cDNAs for human TGF-α, amphiregulin, epiregulin, HB-EGF, EGF, and betacellulin into the 3' end of human placental AP on a pRc/CMV-based mammalian expression plasmid pAlPh *(14)*. This vector contains an N-terminally located HB-EGF signal sequence for proper transport and insertion of the fusion proteins to the membrane. It should be noted that relcase of the AP module into the cell supernatant requires C-terminal (membrane proximal) cleavage of the transmembrane precursor as the N-terminal (membrane distal) clcavage site is destroyed by AP insertion. Although our studies have identified ADAM10 and ADAM17 as membrane-proximal-site-processing enzymes, little information is available on processing at the membrane distal site. EGFR ligand-AP fusion proteins with abolished C-terminal and preserved N-terminal cleavage sites can in theory be used in future studies in an effort to uncover the identities of membrane distal site processing enzymes. Our results obtained with AP-tagged EGFR ligands corroborate previous results that TGF-α, HB-EGF, and amphiregulin are cleaved by ADAM17 *(14)*. Therefore, at least in the case of membrane proximal site cleavage assays, the AP tag does not appear to interfere with the shedding properties of EGFR ligands. In addition, we have observed that processed EGFR ligand–AP fusion proteins are active in receptor binding and signaling (**ref. *14*** and data not shown).

### 3.1. Preparation of Primary Mouse Embryonic Fibroblasts

1. Set up timed matings to generate embryos with the desired genotype, and check for mating plugs the next day.
2. Sacrifice the pregnant female on E13.5.
3. Remove uterine horns, place into a 10-cm tissue culture dish with 10 mL of PBS++ (PBS$^{Ca-Mg}$).
4. Release embryos into PBS; keep on ice from here on.
5. Transfer embryos together with placenta and fetal membrane to fresh PBS++.
6. Remove each embryo from placenta, decapitate with a sharp scalpel, and then dissect out internal organs such as the liver, heart, lung, and intestines.
7. If genotyping is necessary, save the head for DNA isolation.
8. Wash remaining carcass once with PBS–(PBS$^{Ca-Mg}$).
9. Transfer embryo into a tissue culture dish with 2 mL of 0.25% trypsin/EDTA and mince tissue with two sterile scalpels. The efficiency of the mincing greatly determines the yield of mEFs. In order to remove remaining genomic DNA, 1 µL of sterile DNAse can be added to the trypsin incubation if necessary.
10. Incubate 5 min at 37°C. Stop trypsinization by adding 8 mL of mEF growth medium. Pipet up and down, first five times with a 10-mL pipet, then five times with a 5-mL pipet and finally five times with a fire-polished Pasteur pipet to further dissociate cell clusters.
11. Plate cells isolated from one embryo on a 150-cm² tissue culture dish.
12. Change medium the next day. Let cells grow to confluency. After 2–3 d, one should expect to have a confluent 150-cm² plate with approx $1.8 \times 10^7$ cells from one embryo. At this point, the cells can be split into several tissue culture dishes for amplification purposes or may be used directly for shedding assays. Alternatively, the cells may be frozen for future usage (*see* **Note 2**).

### 3.2. Shedding Assays

A typical shedding experiment involves three major steps:

1. Transfection of cells with AP-tagged EGFR ligands.
2. Recovery of mature ligands shed from cells either constitutively or upon stimulation.
3. Analysis and quantitation of shedding.

### 3.2.1. Transfection of Cells With AP-Tagged EGFR Ligands

In our hands, cationic-liposome-mediated transfection of EGFR ligand–AP constructs yields the best results (*see* **Note 5**). The procedure is quick, efficient, and, unlike mammalian cell electroporation, does not require large amounts of DNA or cells to be transfected. However, expression of transfected cDNAs is only transient, and therefore shedding assays must be performed whenever the highest levels of expression are achieved depending on the type of promoter used.

Day 1: Split mEFs into six-well tissue culture dishes plating $3 \times 10^5$ cells per well; grow overnight in mEF growth medium.

Day 2:

1. Prepare transfection mix in two separate sterile polystyrene tubes:
   Solution A: 2 µg DNA per well of cells in 100 µL of plain DMEMs
   Solution B: 5 µL lipofectamine (shake tube to resuspend lipids just before use) in 100 µL plain DMEM per well of cells.
2. Mix DNA and lipofectamine by pipetting Solution A into Solution B to get 200 µL mix per tube). Let stand at room temperature for 30–45 min.
3. Add 800 µL plain DMEM to the 200 µL DNA/lipofectamine mix.
4. Wash cells once with plain DMEM.
5. Add the 1 mL DNA/lipofectamine mixture prepared in **step 3** to cells (1 mL per well); incubate at 37°C for 5 h (*see* **Note 6**).
6. Remove transfection mix, replace with 2 mL of mEF growth medium. Incubate for 24–48 h to achieve maximum levels of expression (*see* **Note 7**).

### 3.2.2. Recovery of Mature Ligands Shed From Cells Either Constitutively or Upon Stimulation

All EGFR ligands are shed from cell membranes constitutively at basal levels; however, in many cases shedding can be upregulated by exogenous stimulants. The phorbol ester PMA is the most commonly used stimulator of ectodomain shedding, yet other stimulants including activators of G protein-coupled receptors such as thrombin, bombesin, carbachol, lysophosphatidic acid (LPA), as well as the tyrosine kinase inhibitor pervanadate or calcium ionophores have also been used for this purpose. As outlined in the introduction, recovery of shed ligands is a two-step process. In the first step, cells in a given tissue culture well are allowed to shed constitutive levels of a particular ligand into conditioned medium for 1 h. In the second step, the same cells are treated for an additional hour with either a stimulator of shedding such as PMA or a suppressor of shedding such as the metalloproteinase inhibitor batimastat (BB94). Upon treatment with the stimulator, shedding of certain EGFR ligands is upregulated; such upregulation over the basal level can be calculated in terms of percentage increase over basal shedding (*see* **Fig. 1B**). Note that the basal level of shedding is evaluated by quantitating the amount of ligand released into cell medium over the first hour under unstimulated conditions (*see* **Note 8**). On the other hand, upon treatment with a metalloproteinase inhibitor such as BB94, shedding of most EGFR ligands is strongly decreased compared to the basal level of shedding. This can be calculated and presented as a percentage decrease in shedding following treatment with the inhibitor and represents the metalloproteinase-dependent component of constitutive EGFR ligand shedding. The logic behind calculating percent suppression of shedding upon treatment with an inhibitor is to evaluate the presence or absence of shedding activities that are sensitive to this inhibitor in different knockout cells lacking a particular sheddase (*see* **Note 9**); for example, our analyses revealed that batimastat-sensitive shedding of EGF and betacellulin was abolished in *adam10–/–* cells, suggesting that ADAM10 is the major batimastat-sensitive component of constitutive EGF and betacellulin shedding (**Note 10**).

3.2.2.1. Collection of Shed Ligands—1st Hour: Basal Shedding

1. At this point, mEFs have already been plated in six-well tissue culture dishes and transfected with a particular EGFR ligand-AP construct 24 h prior to collection.
2. Remove cell medium, wash cells once with Opti-MEM (Gibco) or plain DMEM. From now on, it is important to use serum-free (or reduced-serum) medium as serum components may affect the shedding behavior. Opti-MEM works best in our hands.
3. Add 1 mL of Opti-MEM per well of cells, incubate for 1 h at 37°C.
4. Remove conditioned cell medium into a microfuge tube, keep on ice, and proceed to **Subheading 3.2.2.2., step 1**. These samples are referred to as unstimulated cell supernatants.

3.2.2.2. Collection of Shed Ligands—2nd Hour: Stimulated (or Inhibited) Shedding

1. Dilute stimulant (PMA, discussed previously) or inhibitor (i.e., batimastat, BB94) at the appropriate final concentration into Opti-MEM, add 1 mL per well of cells, incubate for 1 h at 37°C (*see* **Note 11**).
2. Remove conditioned cell medium into a microfuge tube; keep on ice. These samples are referred to as stimulated or inhibited cell supernatants.
3. Lyse cells by adding 0.5 mL of lysis buffer per well; shake at 4°C for 15–30 min (*see* **Note 12**).
4. Transfer cell lysates into microfuge tubes; keep on ice.

5. Clear cell supernatants (unstimulated and stimulated) as well as cell lysates by centrifugation for 30 min at 15,000$g$ in a Sorvall tabletop centrifuge (*see* **Note 8** for a description of how to remove precursor proteins from the culture supernatant, if necessary). Transfer cleared supernatants and lysates into new microfuge tubes without disturbing the pelleted cell debris (*see* **Note 13**). At this point samples can be frozen or analyzed directly.

### 3.2.3. Analysis and Quantitation of Shedding

Shed EGFR ligands in cell supernatants can be detected either through spectrophotometric analysis of AP activity or by running the concentrated supernatants on sodium dodecyl sulfate (SDS)-polyacrylamide gels and staining the gels for AP activity. Spectrophotometric analysis is fast and sensitive, usually 100 µL of 1 mL conditioned cell medium is sufficient for detection. On the other hand, the AP-tagged EGFR ligands in conditioned medium need to be concentrated prior to loading on a gel. Since the alkaline phosphatase moiety is *N*-glycosylated, the lectin ConA can be used as a rapid and efficient means of concentrating the shed fusion proteins.

#### 3.2.3.1. SPECTROPHOTOMETRIC DETECTION OF SHED EGFR LIGANDS

1. Add 100 µL of cleared unstimulated and stimulated cell supernatants to separate wells of a 96-well microplate. Save the rest for SDS-polyacrylamide gel electrophoresis (PAGE) analysis (*see* **Note 14**).
2. Mix with 100 µL of 2 mg/mL 4-nitrophenylphosphate (4-NPP) per well (*see* **Note 15**). 4-NPP is a substrate of alkaline phosphatase, which converts it into nitrophenol, resulting in increased absorbance at 405 nm.
3. Incubate the supernatant–substrate mix at 37°C for color development (yellow). Because it is important to measure $OD_{405}$ while it is still in the linear range, the plates should be visually monitored for color development, and readings in an Elisa plate reader should be taken at different time intervals (every hour over 5–6 h). Often, the plate must be incubated overnight to obtain a good signal-to-noise ratio. Supernatant from nontransfected cells incubated with 4-NPP for the same amount of time is used as the spectrophotometric blank. The absorbance at 405 nm correlates with the amount of shed EGFR ligand in cell supernatant (conditioned medium). Thus, one can calculate the percent increase/decrease in shedding over basal levels upon stimulation/inhibition after measuring the amounts of shed ligands in arbitrary units in stimulated and unstimulated cell supernatants.

#### 3.2.3.2. VISUALIZATION OF SHED EGFR LIGANDS ON SDS-POLYACRYLAMIDE GELS

1. Add 40 µL of ConA resin (equilibrated and resuspended 1:1 in PBS) to the remainder of the cell supernatants (900 µL, after removal of 100 µL for colorimetric AP assay, see above).
2. Incubate at 4°C rotating for 2 h.

3. Spin down ConA beads (at 3000 rpm in a Sorval tabletop centrifuge with adjustable speed for 1 min) and remove supernatant as completely as possible.

4. Elute glycoproteins from the beads by adding 20 μL of elution buffer, mix and then incubate at 37°C for 2 h. Under these conditions, the excess α-D-methylmannoside competes with glycoproteins in the cell supernatant for ConA binding, providing an efficient means of elution without having to boil the samples in SDS–sample loading buffer, which would denature and inactivate the AP module irreversibly (*see* **step 6**).

5. Add 5 μL of 5X SDS-sample loading buffer containing 25 m$M$ dithiothreitol (DTT). Do not boil the samples as AP is irreversibly inactivated at high temperatures (>70°C). Spin down the beads, load supernatants directly on SDS-polyacrylamide gels and apply no more than 100 V to avoid excess heating of the gel during electrophoresis.

6. When the separation is complete, remove gels and incubate them twice for 30 min in 2.5% Triton X-100, followed by a 10-min incubation in alkaline phosphatase buffer (*see* **Note 16**).

7. Visualize the AP activity by adding detection buffer containing the substrate, NBT/BCIP (Roche).

8. Incubate at 37°C until the color development has reached the desired intensity. The enzyme reaction is then stopped in 50% methanol–10% glacial acetic acid (*see* **Note 17**).

## 4. Notes

1. The term primary embryonic fibroblasts or mEF, frequently used in the literature, actually refers to cell preparations that contain a variety of other cell types besides fibroblasts. The main advantage of using primary cells is that the composition of cells should be very similar from experiment to experiment as long as cells are prepared under otherwise identical conditions from different knockout mice.

2. Primary mEFs, like all other primary cells, can be kept in cell culture for a limited number of passages. By the time the shedding assays are performed, the cells should have been passaged no more than 3 times. After passage three the efficiency of transfection is greatly diminished. However, mEFs can be passaged at least twice before performing shedding assays. This is for amplification purposes, because the number of freshly isolated cells (at passage 0) is usually not sufficient to obtain several data points in a given shedding experiment.

3. Every time mEFs lacking a particular gene (i.e., ADAM17) are used for shedding experiments, wild-type mEFs (preferably isolated from wild-type embryos of the same littermate as the knockout ones) should be included as experimental controls. However, if a given knockout mouse is fertile, entire litters with the same genotype can be generated. In this case, separate matings with wild-type mice should be set up at the same time. Nevertheless, the shedding results obtained with the single-well shedding assay using mEFs isolated at different times also produce highly reproducible results.

4. Embryos at E13.5 give the best yield when primary fibroblasts are isolated. Embryos at earlier stages of development are usually too small to generate a sufficient number of cells, and contamination with other differentiated cell types is usually a problem when embryos at later stages of development are used.

5. The method of cDNA transfection should be determined on a case-by-case basis depending on the cell type and cDNA constructs used. For example, for constructs that are not efficiently expressed (i.e., weak promoter, low copy number), electroporation might be preferable. Different types of cationic lipid reagents (i.e., lipofectamine, lipofectamine 2000, fugene) are available. Each has different transfection efficiencies depending on the cell type and density. The expression of transfected ligands should be analyzed by Western blotting or, in the case of AP-tagged EGFR ligands, by running the cell lysate on SDS-polyacrylamide gels and staining the gels for AP activity.

6. Lipofectamine reagent can be toxic to the cells; therefore, transfection reactions should not be allowed to proceed for more than 5 h when primary mEFs are used.

7. Once EGFR ligands are transfected into cells, sufficient levels of expression are usually achieved within 24 h from cytomegalovirus (CMV)-based promoters. However, the time point where expression is appropriate needs to be established on a case-by-case basis for different constructs. Generally, it is critical to wait for the same amount of time before performing the shedding assay to obtain consistent results in different experiments.

8. We perform shedding assays in the presence of stimuli for a time period of 1 h. This condition was chosen because longer exposure of cells to PMA has a number of undesired pleiotropic effects that can affect the interpretation of the results, including stimulation of gene expression, protein synthesis, and intracellular transport. Ectodomain shedding, on the other hand, can be stimulated by PMA within minutes. Collecting cell supernatants for 1 h represents a good compromise between optimizing the amount of shed AP-tagged proteins in the supernatant, while avoiding any undesired side effects of PMA stimulation. The fact that constitutive and stimulated shedding is strongly reduced by addition of BB94 confirms that this assay measures release of EGFR ligands into the supernatant by ectodomain shedding, rather than by increased vesicle shedding, for example. Conversely, if full-length AP-tagged precursor proteins are found in the supernatant sample, this is due to release of membrane fragments from cells during the experiment or harvesting of the sample instead of membrane shedding. These membrane fragments or vesicles containing precursor proteins can be removed from the supernatant by centrifugation at 100,000$g$ for 1 h, which pellets membrane fragments and vesicles containing the precursor protein without noticeably affecting the levels of the solubilized shed ectodomains in the supernatant. However, since precursor proteins usually migrate more slowly than the shed ectodomain—and to avoid interfering with the interpretation of shedding results—we used this additional centrifugation step only to generate publication quality figures, if necessary.

9. Most ectodomain shedding events are mediated by metalloproteinases; therefore, constitutive shedding of many transmembrane proteins including EGFR ligands will be sensitive to treatment with a hydroxamic-acid-type metalloproteinase inhibitor such as batimastat (BB94). However, since batimastat is not commercially available, it can be substituted with other hydroxamates such as TAPI-2 (Peptides International INH-3852). Similarly, cysteine-, aspartyl-, or serine-proteinase-dependent components of constitutive shedding of a shed protein can be analyzed by treating the cells with specific inhibitors of these proteinases.

10. For proper interpretation of the ratiometric results obtained with the AP-shedding assay described here, it is critical to confirm that the absolute levels of constitutive shedding into the cell supernatant of cells isolated from different knockout mice is comparable for any given ligand. If it is not, then ratiometric results may be misleading. To better illustrate this point, assume that a given EGFR ligand has multiple sheddases, two of which are batimastat sensitive and one of which is not. If, hypothetically, one of the batimastat-sensitive activities contributes 50% of ligand shedding, the second contributes 30%, whereas the remaining 20% is not batimastat sensitive, then addition of batimastat to wild-type cells would reduce the constitutive activity by approx 80%. However, if the major batimastat-sensitive activity (50%) is removed in a knockout cell line, then addition of batimastat to these knockout cells would decrease the remaining constitutive activity by 60% compared to untreated cells. This might be difficult to distinguish from an approx 80% reduction after batimastat addition to wild-type cells. However, if absolute shedding levels are compared, then one would be able to identify a reduction of 50% in the absence of the major sheddase. A related issue is whether or not the transfection levels of the precursor protein in the cell lysate should be used as a reference point for normalization of the ratiometric data. In general, it is good practice to independently confirm that the transfection levels as well as the levels of constitutive shedding (discussed previously) are similar in each well of each experiment. However, one caveat of using the AP activity in the cell lysate as a reference is that if stimulation of shedding results in significant consumption of the precursor, then calculating the ratio of protein shed into the supernatant to the remaining precursor in the cell lysate might result in overestimation of the apparent shedding levels. On the other hand, if only a small amount of the total precursor protein is consumed by shedding when cells are stimulated, or if inhibition of the sheddase does not lead to significant accumulation of the precursor (which can be determined in separate experiments), then the AP activity in the cell lysate can be used as a reference point for normalization in shedding experiments.

11. The final PMA concentration should be 20 ng/mL. We prepare a stock solution of 5 mg/mL in ethanol, which is kept at –80°C. From this 250,000× master stock, small aliquots of 1000× stocks (further diluted in ethanol) can be prepared and kept at –20°C. PMA should never be allowed to stay at room temperature. Note that it is also light sensitive. We store batimastat (BB94) at –80°C and use at a final concentration of 1 μ*M*.

12. When lysing the cells, proteinase inhibitors must be included in the lysis buffer. Lysis should be carried out at 4°C, and cell lysates must be kept on ice or frozen immediately to prevent protein degradation.

13. The cell lysates should be analyzed to confirm the expression of transfected cDNAs. They can also be used to confirm equal expression levels in different cell types and experiments.

14. When analyzing the AP levels in cell supernatants spectrophotometrically, averages of duplicate or triplicate readings from a given sample provide more reliable results.

15. Prepare a 50 mg/mL stock solution of 4-NPP and keep at 4°C, protected from light. For spectrophotometric analysis of AP activity, dilute the 4-NPP stock 1:25 to 2 mg/mL in 100 m$M$ Tris (pH 9.5), 100 m$M$ NaCl, 20 m$M$ MgCl$_2$.

16. Gels should not be washed for longer periods than suggested, because this will cause the AP signal to diffuse, thereby creating fuzzy bands.

17. AP is active under slightly basic conditions; thus the reaction is performed in 100 m$M$ Tris (pH 9.5), 100 m$M$ NaCl, 20 m$M$ MgCl$_2$. Addition of 50% methanol–10% glacial acetic acid will not only stop the AP reaction by lowering the pH but also fix the proteins in the gels, creating sharper bands.

## References

1. Yarden, Y. and Sliwkowski, M. X. (2001) Untangling the ErbB signaling network. *Nat. Rev. Mol. Cell. Biol.* **2,** 127–137.
2. Gschwind, A., Fischer, O. M., and Ullrich, A. (2004) The discovery of receptor tyrosine kinases: targets for cancer therapy. *Nat. Rev. Cancer* **4,** 361–370.
3. Schlessinger, J. (2002) Ligand-induced, receptor-mediated dimerization and activation of EGF receptor. *Cell* **110,** 669–672.
4. Burgess, A. W., Cho, H. S., Eigenbrot, C., et al. (2003) An open-and-shut case? Recent insights into the activation of EGF/ErbB receptors. *Mol. Cell.* **12,** 541–552.
5. Harris, R. C., Chung, E., and Coffey, R. J. (2003) EGF receptor ligands. *Exp. Cell. Res.* **284,** 2–13.
6. Brachmann, R., Lindquist, P. B., Nagashima, M., et al. (1989) Transmembrane TGF-alpha precursors activate EGF/TGF-alpha receptors. *Cell* **56,** 691–700.
7. Wong, S. T., Winchell, L. F., McCune, B. K., et al. (1989) The TGF-alpha precursor expressed on the cell surface binds to the EGF receptor on adjacent cells, leading to signal transduction. *Cell* **56,** 495–506.
8. Blobel, C. P. (2004) ADAMs: key players in EGFR-signaling, development and disease. *Nat. Rev. Mol. Cell Biol.* **6,** 32–43.
9. Dong, J., Opresko, L. K., Dempsey, P. J., Lauffenburger, D. A., Coffey, R. J., and Wiley, H. S. (1999) Metalloprotease-mediated ligand release regulates autocrine signaling through the epidermal growth factor receptor. *Proc. Natl. Acad. Sci. USA* **96,** 6235–6240.
10. Prenzel, N., Zwick, E., Daub, H., et al. (1999) EGF receptor transactivation by G-protein-coupled receptors requires metalloproteinase cleavage of proHB-EGF. *Nature* **402,** 884–888.

11. Schafer, B., Gschwind, A., and Ullrich, A. (2004) Multiple G-protein-coupled receptor signals converge on the epidermal growth factor receptor to promote migration and invasion. *Oncogene* **23,** 991–999.

12. Peschon, J. J., Slack, J. L., Reddy, P., et al. (1998) An essential role for ectodomain shedding in mammalian development. *Science* **282,** 1281–1284.

13. Jackson, L. F., Qiu, T. H., Sunnarborg, S. W., et al. (2003) Defective valvulogenesis in HB-EGF and TACE-null mice is associated with aberrant BMP signaling. *EMBO J.* **22,** 2704–2716.

14. Sahin, U., Weskamp, G., Zhou, H. M., et al. (2004) Distinct roles for ADAM10 and ADAM17 in ectodomain shedding of six EGFR-ligands. *J. Cell. Biol.* **164,** 769–779.

15. Zheng, Y., Schlondorff, J., and Blobel, C. P. (2002) Evidence for regulation of the tumor necrosis factor alpha-convertase (TACE) by protein-tyrosine phosphatase PTPH1. *J. Biol. Chem.* **277,** 42,463–42,470.

16. Hartmann, D., de Strooper, B., Serneels, L., et al. (2002) The disintegrin/metalloprotease ADAM 10 is essential for Notch signaling but not for alpha-secretase activity in fibroblasts. *Hum. Mol. Genet.* **11,** 2615–2624.

# 8

## In Vitro and In Vivo Assays of Monoubiquitination of Receptor Tyrosine Kinases

### Menachem Katz, Yaron Mosesson, and Yosef Yarden

### Summary

Growth factor receptors, such as the epidermal growth factor receptor (EGFR), stimulate a variety of signal transduction pathways upon binding a ligand molecule at the cell surface. Desensitization of signaling initiates when active receptors are recruited to clathrin-coated regions of the plasma membrane and subsequently sorted to intracellular degradation in lysosomes. Sorting for lysosomal degradation entails receptor conjugation with ubiquitin molecules, which are recognized by the endocytic machinery. Unlike degradation in the 26S proteasome, which requires a chain of four or more units of ubiquitin (polyubiquitination), covalent addition of a monomeric ubiquitin (monoubiquitination) appears sufficient for receptor sorting to lysosomal degradation. In this chapter we describe two methods that contrast polyubiquitination with monoubiquitination of EGFR. Because monoubiquitination enables evasion from proteasomal degradation, the methods we describe may be useful for the analysis of other monoubiquitination events.

**Key Words:** E3 ubiquitin ligase; epidermal growth factor; endocytosis; ErbB/HER; signal transduction; tyrosine kinase; ubiquitin.

## 1. Introduction

Protein ubiquitination refers to the covalent conjugation of the 76-amino-acid polypeptide ubiquitin to substrate proteins. This process involves the sequential transfer of the ubiquitin moiety through three enzymes: ubiquitin-activating enzyme (E1); ubiquitin-conjugating enzyme (E2); and ubiquitin ligase (E3), which is important for the specific recognition of the substrate. The E3 either accepts ubiquitin and directly transfers it to the substrate or serves as a bridge between the E2 and the substrate (reviewed in **ref. 1**). As a result, an isopeptide bond is formed between the glycine located at the carboxyl-terminus of ubiquitin and the ε-amino group of a lysine in the sub-

From: *Methods in Molecular Biology, vol. 327: Epidermal Growth Factor: Methods and Protocols*
Edited by: T. B. Patel and P. J. Bertics © Humana Press Inc., Totowa, NJ

**A**

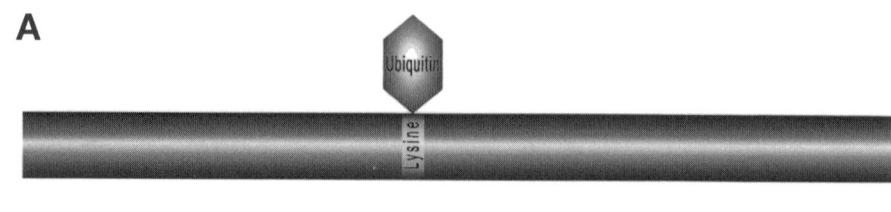

**B**

**C**

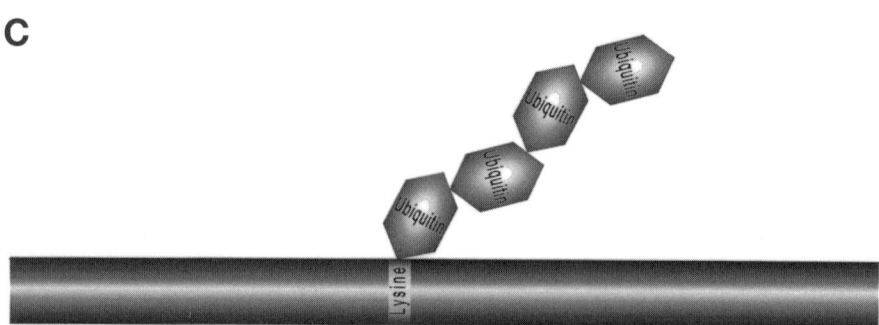

Fig. 1. Distinct patterns of protein ubiquitination. **(A)** Attachment of a single ubiquitin molecule to a lysine residue of the substrate protein (monoubiquitination). **(B)** Attachment of several monoubiquitins to several lysines of the substrate protein (multiubiquitination). **(C)** The addition of a polymeric ubiquitin chain to a lysine of the substrate protein (polyubiquitination).

strate protein. Proteins can be conjugated with ubiquitin in three different modes **(Fig. 1)**: (1) monoubiquitination refers to the addition of one ubiquitin molecule to a lysine of the target protein; (2) multiple monoubiquitination (hereinafter multiubiquitination) indicates the conjugation of several monoubiquitins to several distinct lysines in the acceptor protein; (3) polyubiquitination refers to conjugation of a polymerized ubiquitin to one or several lysines of a substrate. Within the chain of polymeric ubiquitin, the carboxyl-terminal glycine of ubiquitin is linked to a lysine of the preceding ubiquitin. Lysine 48 of ubiquitin commonly serves as the acceptor lysine in the chain, although lysines 11, 29, and 63 were also found to serve as acceptor lysines (reviewed in **ref. 2**).

The pattern of ubiquitination of a given protein may dictate the fate of the protein. Usually proteins that are conjugated with a polyubiquitin chain, at least four ubiquitins long *(3)*, are destined for degradation by the 26S proteasome. On the other hand, membrane proteins that are conjugated with monoubiquitin or a chain of diubiquitin are usually sorted for endocytosis and degradation in lysosomes (reviewed in **ref. 4**). The plasma membrane receptor of the epidermal growth factor (EGFR, also known as ErbB-1) is one of the most extensively characterized receptor tyrosine kinases. EGFR undergoes internalization and sorting to the lysosome, following the binding of EGF or other ligands. This process is accompanied by the ubiquitination of the receptor, which is mediated by the E3 ubiquitin ligase c-Cbl *(5,6)*. Since both poly-ubiquitination and mono-/multiubiquitination may serve as signals for protein degradation, either by the proteasome or by the lysosome, respectively, elucidating the pattern of ubiquitination is important for understanding the process of receptor degradation. (The reader is also encouraged to refer to Chapter 9 regarding discussion of additional concepts and methods related to this subject.)

Ubiquitylation of substrate proteins, including EGFR, usually manifests as a smeary appearance of protein bands when separated by polyacrylamide gel electrophoresis. However, polyubiquitination and multiubiquitination give rise to indistinguishable electrophoretic patterns, which calls for alternative analytical methods. The assays detailed below, as well as additional tests, enabled identification of EGFR ubiquitination, following stimulation by EGF, as multi- rather than polyubiquitination *(7,8)*. The simplest approach used to identify the type of ubiquitin modification of the EGFR is by immunoprecipitation of the receptors from cells followed by Western blotting utilizing specific antiubiquitin antibodies. These antibodies can be used to differentiate between receptor polyubiquitination and receptor multiubiquitination *(7)*. The immunological technique, however, is limited to cells in which the expression of both ubiquitin and receptors is substantially high. Alternatively, ectopically expressed peptide-tagged forms of mutant ubiquitin molecules can be used. Expression of an ubiquitin mutant impaired in its ability to form polymeric chains (primarily an ubiquitin whose lysine 48 is replaced by an alanine) will allow one to distinguish between poly- and multiubiquitylation. However, the interpretation of experiments based on expression of mutant ubiquitin molecules is complicated, since cells endogenously express wild-type ubiquitin, which may form mixed ubiquitin chains composed of wild-type and mutant ubiquitins.

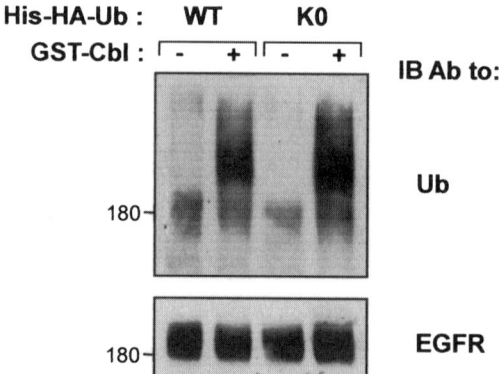

Fig. 2. Multiple monoubiquitination of EGFR in vitro. EGFR immunoprecipitates from A431 cells were subjected to ubiquitination in vitro in the presence of reticulocyte lysate, together with either wild-type or lysine-less ubiquitin produced in bacteria (His-HA-Ub-WT or -KO, respectively). GST-Cbl was added as indicated. Reaction mixtures were incubated at 30°C for 60 min before immunoblotting (IB) with the indicated antibodies. Note that similar patterns of ubiquitination were obtained with the two distinct forms of ubiquitin.

Two alternative methods are used to elucidate the type of protein ubiquitination. One method utilizes mutant ubiquitins in a cell-free system (*see* **Notes 1** and **2**; **Fig. 2**). Isolated receptors are incubated with a rabbit reticulocyte lysate supplemented with a recombinant peptide-tagged ubiquitin, either wild-type or a mutant with all its lysines knocked-out (KO). The ubiquitylated receptors are then subjected to Western blot analysis directed to the tagged ubiquitins. Receptors which are modified with multiple monoubiquitins will appear similarly whether incubated with wild-type or mutated ubiquitin. On the other hand, polyubiquitylated receptors will appear smeared after incubation with wild-type ubiquitin and as a distinct band following incubation with the lysine-less ubiquitin mutant. The other method is based on simultaneous ectopic expression of two different peptide-tagged ubiquitins mutated in the major chain-branching residues (i.e., lysines 11, 29, 48, and 63). Sequential rounds of immunoprecipitation and elution, followed by Western blotting, enable determination of the type of ubiquitination (**Fig. 3**).

## 2. Materials

### *2.1. Equipment*

1. Tissue culture facilities.
2. A refrigerated benchtop centrifuge.
3. Gel electrophoresis apparatus and a power supply.

4. A trans-blot blotting apparatus and a power supply.
5. Heating block (100°C).
6. Film-developing apparatus.

## 2.2. Reagents

1. Phosphate-buffered saline (PBS): 8 g/L NaCl, 0.2 g/L KCl, 1.44 g/L $Na_2HPO_4$, 0.24 g/L $KH_2PO_4$, adjust pH to 7.2–7.4 with HCl.
2. Solubilization buffer: 50 m$M$ hydroxyethyl piperazine ethan sulfonate (HEPES) (pH 7.5), 150 m$M$ NaCl, 10% glycerol (w/v), 1% (w/v) Triton X-100, 1 m$M$ ethylenediaminetetraacetic acid (EDTA), 1 m$M$ ethylene glycol tetraacetic acid (EGTA), 10 m$M$ NaF, 30 m$M$ β-glycerol phosphate, 0.2 m$M$ $Na_3VO_4$ and a proteinase inhibitor cocktail (from Sigma). *Note:* $Na_3VO_4$ and the proteinase inhibitor cocktail are freshly added to the buffer. They are stored in stock solutions at –20°C.
3. HNTG buffer: 20 m$M$ HEPES (pH 7.5), 150 m$M$ NaCl, 0.1% (w/v) Triton X-100, and 10% (w/v) glycerol.
4. RIPA buffer: 25 m$M$ Tris-HCl (pH 7.5), 150 m$M$ NaCl, 0.5% Na–deoxycholate, 1% (w/v) NP-40, and 0.1% (w/v) sodium dodecyl sulfate (SDS).
5. TBST buffer: 20 m$M$ Tris-HCl (pH 7.5), 150 m$M$ NaCl, 0.05% (w/v) Tween 20.
6. Antibodies: An anti-EGFR monoclonal antibody suitable for immunoprecipitation and commercial antibodies to the hemagglutinin (HA) and Flag peptides. Secondary antibodies conjugated to horseradish peroxidase (HRP) are commercially available.
7. Protein gel sample buffer (sixfold concentrated): 350 m$M$ Tris-HCl (pH 8.8), 30% glycerol, 10% SDS, 9.3% dithiothreitol (DTT), and 0.12 mg/mL bromophenol blue. The solution is kept at –20°C in aliquots and thawed at 37°C to dissolve the SDS.
8. Coomassie stain: 0.25% (w/v) Coomassie brilliant blue in 50% (v/v) methanol.
9. Nitrocellulose membranes.
10. Enhanced chemiluminescence (ECL) reagent.
11. X-ray films.
12. Bradford protein-detection reagent.
13. Blocking solution: 5% low-fat milk in TBST buffer.
14. Stripping buffer: 100 m$M$ 2-mercaptoethanol, 2% SDS, 62.5 m$M$ Tris-HCl (pH 6.7).
15. Bacterial expression vectors: c-Cbl cloned into pGEX vector (Pharmacia Biotech); HA-tagged ubiquitin, either wild-type or a lysine-less mutant (KO), cloned into the pET28 plasmid (Novagen). *Note:* The pET28 vector itself adds a polyhistidine peptide tag.
16. Bacterial strains: BL21 (DE3) *Escherichia coli* bacteria (Invitrogen); Rosetta DE3 pLysS *E. coli* bacteria (Novagen).
17. Antibiotics: ampicillin (100 mg/mL) and kanamycin (50 mg/mL).
18. 2× YT medium: tryptone (16 g/L), yeast extract (10 g/L), NaCl (15g/L); adjust to pH 7.0 with NaOH.
19. 100 m$M$ isopropyl-β-D-thio-galactopyranoside (IPTG)
20. Triton X-100, 20% (w/v) stock solution.

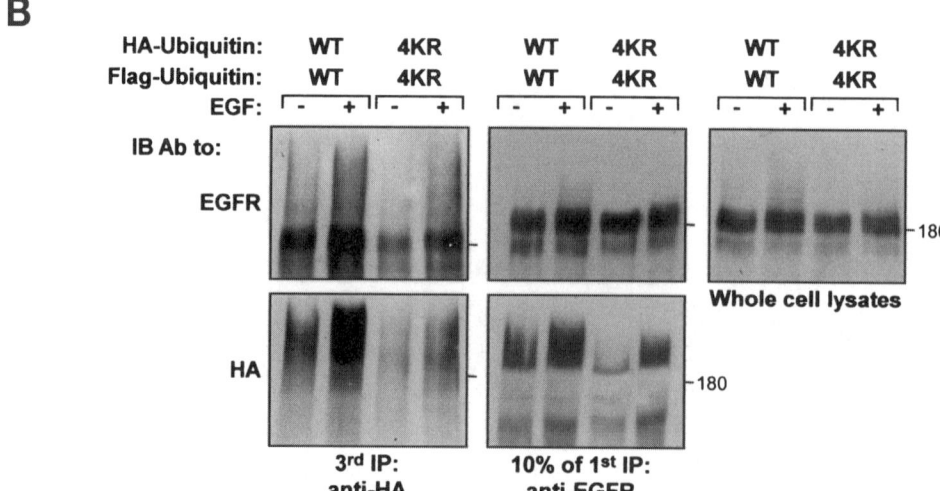

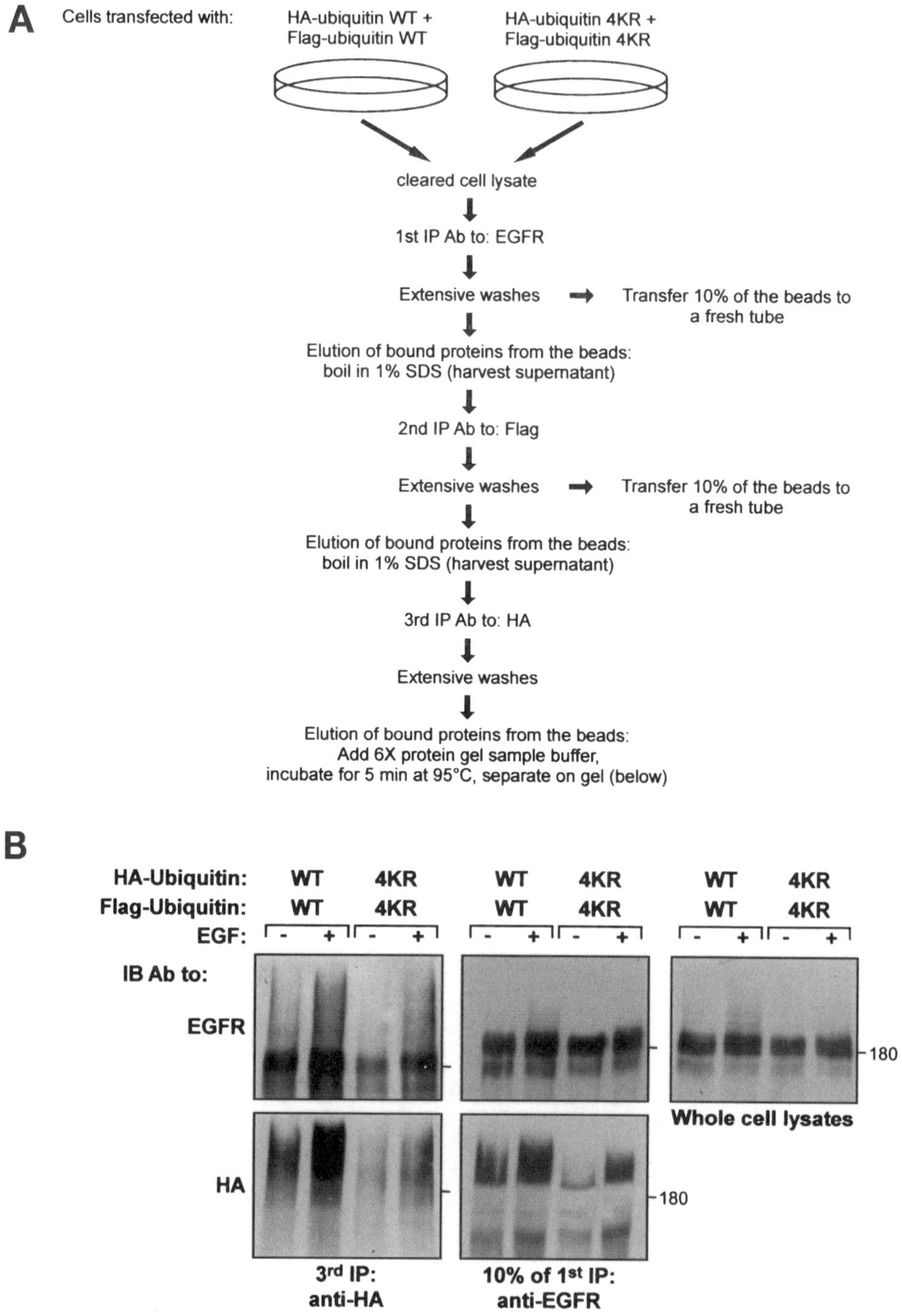

**A**  Cells transfected with:

HA-ubiquitin WT +
Flag-ubiquitin WT

HA-ubiquitin 4KR +
Flag-ubiquitin 4KR

cleared cell lysate

1st IP Ab to: EGFR

Extensive washes  →  Transfer 10% of the beads to
a fresh tube

Elution of bound proteins from the beads:
boil in 1% SDS (harvest supernatant)

2nd IP Ab to: Flag

Extensive washes  →  Transfer 10% of the beads to
a fresh tube

Elution of bound proteins from the beads:
boil in 1% SDS (harvest supernatant)

3rd IP Ab to: HA

Extensive washes

Elution of bound proteins from the beads:
Add 6X protein gel sample buffer,
incubate for 5 min at 95°C, separate on gel (below)

**B**

| HA-Ubiquitin: | WT | 4KR | WT | 4KR | WT | 4KR |
| Flag-Ubiquitin: | WT | 4KR | WT | 4KR | WT | 4KR |

EGF:  - +  - +    - +  - +    - +  - +

IB Ab to:

EGFR

HA

3rd IP:
anti-HA

10% of 1st IP:
anti-EGFR

Whole cell lysates

180
180

21. Dialysis bag with a 6000–8000 Da nominal molecular weight cut-off.
22. Glutathione elution buffer: 7.5 m$M$ glutathione, 50 m$M$ Tris-HCl (pH 8.0).
23. Agarose beads: Ni$^{2+}$-conjugated agarose beads; anti-mouse IgG coupled to agarose beads; anti-HA antibodies coupled to agarose beads; anti-FLAG antibodies coupled to agarose beads. Glutathione-coupled agarose beads.
24. His-wash buffer: 50 m$M$ NaH$_2$PO$_4$, 300 m$M$ NaCl, 20 m$M$ imidazole, 15% (w/v) glycerol. Adjust pH to 8.0 with NaOH.
25. His-lysis buffer: 50 m$M$ NaH$_2$PO$_4$, 300 m$M$ NaCl, 15% glycerol, 10 m$M$ imidazole, 1% Triton X-100, fresh bacterial proteinase inhibitors cocktail (amount as suggested by the manufacturer); adjust pH to 8.0 with NaOH.
26. His-elution buffer: 50 m$M$ NaH$_2$PO$_4$, 300 m$M$ NaCl, 250 m$M$ imidazole, adjust pH to 8.0 with NaOH.
27. Rabbit reticulocyte lysate (from a commercial source).
28. In vitro wash buffer: 40 m$M$ Tris-HCl (pH 7.5), 5 m$M$ MgCl$_2$, 2 m$M$ DTT.
29. In vitro ubiquitination buffer: 40 m$M$ Tris-HCl (pH 7.5), 5 m$M$ MgCl$_2$, 2 m$M$ dithiothreitol, 2 m$M$ ATP, 5 µL rabbit reticulocyte lysate, and 0.2 µg GST-Cbl protein, per reaction. The enzymes and ATP are unstable in this solution. Hence it is recommended to prepare this buffer fresh prior to each experiment.
30. Cell growth medium: Dulbecco's modified Eagle's medium (DMEM) supplemented with 10% fetal calf serum (FCS), penicillin (100 U/mL), streptomycin (0.1 mg/mL), and pyruvate (1 m$M$).
31. Cell lines: confluent 100-mm plates of A431 cells (these cells express high levels of EGFR) and similar plates of HEK-293T cells ready for transfection (80% confluence).
32. Ca$^{2+}$-phosphate transfection reagent: 1.25 $M$ CaCl$_2$; 2X HBS: 280 m$M$ NaCl, 10 m$M$ KCl, 1.5 m$M$ Na$_2$HPO$_4$, 12 m$M$ dextrose, 50 m$M$ HEPES (pH 7.05). Sterilize the solution after preparation, and keep it sterile.

---

Fig. 3. *(opposite page)* Multiple monoubiquitination of epidermal growth factor receptor (EGFR) in living cells. (**A**) Scheme of the experimental procedure. (**B**) HEK-293T cells transfected with vectors encoding EGFR and c-Cbl together with a mixture of two plasmids encoding Flag-ubiquitin and HA-ubiquitin, either wild-type or 4KR (lysines 11, 29, 48, and 63 are substituted with arginines). Forty-eight hours after transfection, cells were incubated at 37°C for 10 min without or with EGF (100 ng/mL). Cell lysates were subjected to three steps of sequential immunoprecipitation (IP) with the following antibodies: an antiEGFR antibody, an anti-Flag antibody, and an anti-HA antibody (*see* **A**), followed by immunoblotting (IB) with anti-EGFR or anti-HA antibodies, as indicated (**left column**). The middle column shows the respective immunoblotting of 10% of the first immunoprecipitation. The right panel shows immunoblotting of whole cell lysates with an anti-EGFR antibody. Note that an elongation-defective mutant of ubiquitin (4KR) yielded a pattern similar to wild-type ubiquitin, which excludes polyubiquitination of EGFR in living cells. Because successive immunoprecipitation with two antipeptide tags resulted in smeary receptor patterns, it is conceivable that multiple lysines of each receptor are simultaneously monoubiquitylated.

33. Mammalian expression vectors containing the following cDNAs: EGFR; c-Cbl; HA- or Flag-tagged ubiquitin, either wild-type or a mutant whose lysines 11, 29, 48, and 63 were replaced by arginines (4KR).
34. Starvation medium: cell growth medium containing 0.1% FCS.
35. EGF (100 µg/mL).
36. Immunoprecipitation (IP) elution buffer: 2% SDS, 50 m$M$ Tris-HCl (pH 7.5).

## 3. Methods

### 3.1. In Vitro Ubiquitination of EGFR (see Fig. 2 and Notes 1–7)

In this assay, a cell-free system is used to compare the pattern and intensity of receptor conjugation with either wild-type or lysine-less ubiquitin (KO). A recombinant E3 ubiquitin ligase (i.e., c-Cbl) and the wild-type and lysine-less ubiquitins may be prepared in advance and stored at –70°C. The EGFR must be purified on the day of the experiment.

#### 3.1.1. Purification of a GST-c-Cbl Fusion Protein From Bacteria

This protocol follows the standard protocol of the manufacturer (Pharmacia). Because full-length c-Cbl is expressed at low levels in bacteria, it is recommended to use large volumes of bacterial cultures (*see* **Note 3**).

1. Transform BL-21 *E. coli* bacteria with a glutathione-*S*-transferase (GST)-c-Cbl bacterial expression vector (pGEX) using the heat shock method.
2. Inoculate 10 mL of 2X YT medium supplemented with 10 µL ampicillin (to final concentration of 100 µg/mL) using a single colony of BL-21 transformed cells.
3. Incubate the cells overnight at 37°C with vigorous shaking.
4. Transfer the 10 mL culture into pre-warmed 2X YT medium (1 L) supplemented with 1 mL of ampicillin. Incubate the culture at 30°C with shaking (~250 rpm) until the OD at $A_{600}$ reaches 0.4.
5. Add 5 mL of the IPTG solution to final concentration of 0.5 m$M$ and continue the incubation for an additional 4 h.
6. Pellet the bacteria by centrifugation at 2700$g$ for 10 min at 4°C, and resuspend on ice in 50 mL ice-cold PBS. Disrupt bacteria using sonication, followed by incubation for 30 min at 4°C with 1% Triton X100.
7. Centrifuge at 12,000$g$ for 10 min at 4°C, transfer the supernatant to fresh 50-mL conical tubes (divide the lysate into two tubes, 25 mL each).
8. Add to each of the tubes 0.5 mL of a 50% slurry of glutathione-agarose equilibrated with PBS. Incubate for 1 h at 4°C while shaking.
9. Centrifuge the tubes at 1000$g$ for 5 min. Discard supernatant being careful not to resuspend the beads. Add to the tube 10 mL of ice-cold PBS and mix; repeat this step three times.
10. Resuspend the beads in 250 µL of PBS. Combine the beads from the two tubes into a fresh 1.5-mL tube using a trimmed pipet tip.

11. Remove the PBS from the beads after a short spin at 14,000 rpm using a benchtop microcentrifuge. Elute the GST-Cbl fusion protein from the beads by adding 1.2 mL of glutathione elution buffer and incubating the tube at room temperature for 10 min. Collect the eluted proteins into a fresh tube. Repeat this step three times.
12. Transfer 20 µL of each elution fraction into a fresh tube and add 4 µL of the 6X protein gel sample buffer. Make a small hole in the lid of each tube using a needle. Incubate the sample at 95°C for 5 min, then centrifuge briefly at 14,000 rpm.
13. Load the samples onto a 10% acrylamide SDS-polyacrylamide gel electrophoresis (PAGE) minigel using long narrow pipet tips. Run the gel at 100 V for approx 1.5 h (size of GST-Cbl is approx 150 kDa). Stain the gel-resolved proteins by using a Coomassie stain or equivalent gel stain methods in order to determine levels of the eluted GST fusion in the series of fractions.
14. Dialyze the fractions containing high levels of GST-Cbl for 24 h at 4°C against PBS. Replace dialysis buffer at least thrice to reach a $10^6$:1 ratio of PBS to sample. Proteins can be stored in aliquots at –70°C.

### 3.1.2. Purification of His-HA-Tagged Ubiquitin From Bacteria

1. Transform Rosetta DE3 pLysS bacteria with His-HA-tagged ubiquitin (either wild-type or KO mutant) bacterial expression vector (cloned into the pET vector) using the heat shock method.
2. Inoculate 4 mL of 2X YT medium supplemented with 4 µL of kanamycin (to final concentration of 50 µg/mL) using a single colony of the Rosetta transformed cells.
3. Incubate the cells overnight at 37°C with vigorous shaking.
4. Transfer 4 mL of the culture into 400 mL of prewarmed 2X YT medium supplemented with 400 µL of kanamycin. Incubate the culture at 30°C with shaking (~250 rpm) until an OD of 0.5 at $A_{600}$ is reached.
5. Add 2 mL of the IPTG solution to final concentration of 0.5 m$M$, and continue the incubation for an additional 4 h.
6. Weigh empty centrifuge buckets. Transfer the culture into the buckets and pellet the bacteria by centrifugation at 2700$g$ for 10 min at 4°C. Drain the pellet and weigh the bucket again to estimate the weight of bacterial pellet.
7. Resuspend the pellet, on ice, in 3 mL of ice-cold His-lysis buffer per gram pellet.
8. Disrupt bacteria using sonication.
9. Centrifuge at 12,000$g$ for 20 min at 4°C and transfer the supernatant to a fresh tube.
10. Add to 4 mL of clear lysate a slurry (50%) of $Ni^{2+}$-conjugated agarose beads (1 mL). Incubate for 1 h at 4°C while rotating.
11. Centrifuge the tubes at 1000$g$ for 5 min. Discard supernatant while being careful not to resuspend the beads. Add to the tube 10 mL of ice-cold His-wash buffer, repeat this step three times.
12. Elute the His-HA-tagged ubiquitin from the beads by adding His-elution buffer in a volume equivalent to the bead bed volume. Incubate for 20 min at room temperature with shaking.

13. Centrifuge for a short spin at 14,000 rpm, collect the supernatant into a fresh tube.
14. Dialyze the His-HA-ubiquitin eluate for 24 h at 4°C against PBS using low molecular weight cut-off dialysis tubing (discussed previously).
15. Transfer 20 µL of dialyzed protein into a fresh tube and add 4 µL of 6X protein gel sample buffer. Make a small hole in the lid of each tube, using a needle. Incubate the sample at 95°C for 5 min, then centrifuge briefly at 14,000 rpm.
16. Run the sample on a SDS-PAGE minigel (15% acrylamide) and Coomassie stain in order to verify the purity of the prepared ubiquitin. The size of His-HA-ubiquitin is around 10 kDa. Thereafter, protein concentration should be estimated by using the Bradford method. Proteins can be stored in aliquots at –70°C.

### 3.1.3. Immunoprecipitation of EGFR From A431 Cells

For each in vitro ubiquitination reaction receptors are produced from one confluent 100-mm plate of A431 cells.

1. Remove 20 µL of 50% slurry of anti-mouse IgG-agarose beads per reaction into a tube and wash twice with 500 µL of solubilization buffer.
2. Resuspend the beads in approx 300 µL of solubilization buffer and add 2 µg of an anti-EGFR mouse monoclonal antibody per reaction.
3. Incubate the beads with gentle agitation at 4°C for 1 h to allow coupling of the anti-EGFR antibodies.
4. Wash the beads twice with 1 mL of solubilization buffer and resuspend in 50 µL of solubilization buffer per reaction.
5. Place a confluent 100-mm plate of A431 cells on ice. Wash the cell monolayer once with 5 mL of ice-cold PBS.
6. Add to the plate 1 mL of ice cold solubilization buffer. Collect the lysate in a prechilled 1.5-mL tube using a cell scraper.
7. Vortex the mixture and incubate on ice for 20 min.
8. Spin the tube using a benchtop microcentrifuge at 14,000 rpm for 20 min at 4°C.
9. Transfer the supernatant into a tube containing 50 µL of a slurry of agarose beads precoupled to an anti-EGFR antibody. For several reactions, precoupling (**steps 1–4**) can be carried out in a 15-mL tube, and the cleared lysates can be pooled into this tube.
10. Incubate the tube while gently rotating for one hour at 4°C to allow immunoprecipitation of the receptors.
11. Wash the beads twice with 1 mL of HNTG, and then twice with the in vitro wash buffer. At this stage the receptors are immobilized, purified, and ready for use in an in vitro ubiquitination reaction.

### 3.1.4. In Vitro Receptor Ubiquitination Assay

1. Purify EGFR from A431 cells as described in **Subheading 3.1.3.** If the receptor protein was pooled, aliquot evenly into separate 1.5-mL tubes, and carefully remove residual buffer following brief centrifugation.
2. Thaw aliquots of GST-Cbl (prepared as in **Subheading 3.1.1.**), His-HA-ubiquitins, either wild-type or KO (*see* **Subheading 3.1.2.** and **Note 4**), and the rabbit reticulocyte lysate. Prepare a fresh in vitro ubiquitination buffer.

3. Add 40 µL of in vitro ubiquitination buffer to each reaction tube on ice. Subsequently add 0.5 µg of the purified wild-type or KO ubiquitin.
4. Incubate the tube for 1 h at 30°C with gentle agitation.
5. Wash the beads four times with 1 mL of RIPA buffer.
6. At the end of the last wash, leave approx 40 µL of buffer on top of the beads. Add to the beads 8 µL of 6X protein gel sample buffer.
7. Make a small hole in the lid of each tube, using a needle. Incubate the sample at 95°C for 5 min, then centrifuge for a brief spin at 14,000 rpm.
8. Load the sample onto a 6.5% acrylamide minigel using long narrow pipet tips. Run the gel at 100 V for around 1.5 h (size of EGFR is 170 kDa).
9. Transfer the proteins from the gel onto a nitrocellulose membrane, using 200 mA electrical current (approx 75 min). It is important to keep the transfer buffer cold during this step.
10. Block the membrane by incubating in a blocking solution for 30 min at room temperature with shaking.
11. Remove the blocking solution and wash the membrane three times with TBST buffer. Incubate the membrane with antibody directed either to the HA tag or to ubiquitin (diluted as suggested by the manufacturer) overnight at 4°C with constant shaking.
12. Remove the antibody from the membrane and wash three times (5 min each) with TBST. Add the appropriate secondary antibody conjugated to HRP to the membrane and incubate for 30 min with shaking at room temperature.
13. Wash the membrane three times with TBST. Develop the blot using an ECL kit, according to the manufacturer's instructions. Expose to an x-ray film. The membrane can be further stripped and blotted for the receptor as indicated in **Subheading 3.2.3.**, **step 4** (*see* **Notes 5–7**).

## 3.2. Receptor Ubiquitination in Living Cells Analyzed by Sequential Immunoprecipitation (see Fig. 3)

### 3.2.1. Cell Transfection and Preparation of Cleared Cell Lysates

1. Transfect four plates of HEK-293T cells (*see* **Note 10**) using the $Ca^{2+}$-phosphate method. Two plates are to be transfected with plasmids encoding EGFR and c-Cbl, together with vectors for wild-type ubiquitin conjugated to an HA tag and an ubiquitin conjugated to a Flag tag. The other two plates are transfected with EGFR and c-Cbl plasmids, together with a plasmid-driving expression of ubiquitin-4KR, an ubiquitin mutated in all of four lysines known to form ubiquitin branching (i.e., lysines 11, 29, 48, and 63). Both HA- and Flag-tagged 4KR-ubiquitins are used.
2. Thirty-six hours after transfection, replace the cell medium with 5 mL of starvation medium.
3. Forty-eight hours after transfection, prepare fresh solubilization buffer (it is advised to perform **Subheading 3.2.2.**, **steps 1–3** at this time). Prewarm to 37°C 10 mL of starvation medium and 10 mL of starvation medium containing 10 µL of EGF (final concentration 100 ng/mL).

4. Aspirate the medium from the cells and add starvation medium to one of the plates that was transfected with wild-type ubiquitin and to one of the plates transfected with 4KR-ubiquitin. Add medium containing EGF to the other two plates.
5. Transfer the cells to a 37°C incubator and incubate for 10 minutes.
6. At the end of the incubation, place the cells in an ice bucket. Gently wash the cells with 4 mL of ice-cold PBS (*Note:* HEK-293T cells easily detach from the plate), and add to the plate 1 mL of solubilization buffer.
7. Collect the lysate in a prechilled 1.5-mL tube using a cell scraper. Vortex mix the tube and incubate on ice for 20 min.
8. Spin the tube using a benchtop microcentrifuge at 14,000 rpm for 20 min at 4°C and transfer the supernatant into a fresh tube.
9. Perform a Bradford assay to ensure equal loading of proteins in the immunoprecipitation reaction.

### 3.2.2. Sequential Immunoprecipitation

1. Remove 100 μL of 50% slurry of anti-mouse IgG-agarose beads into a tube and wash twice with 500 μL of solubilization buffer.
2. Resuspend the beads in approx 300 μL of solubilization buffer and add 10 μg of an anti-EGFR mouse monoclonal antibody.
3. Incubate the beads with gentle agitation at 4°C for 1 h to allow coupling of the anti-EGFR antibodies.
4. Wash the beads twice with 1 mL of solubilization buffer and resuspend them in 200 μL of solubilization buffer.
5. Aliquot 50 μL of the beads into four 1.5-mL tubes using a trimmed pipet tip and add equal amounts of total cell lysate (prepared as described under **Subheading 3.2.1.**). *Note:* Take as much as possible from the sample containing the least amount of protein and respective volumes from the other samples.
6. Incubate the tube while gently rotating for 1 h at 4°C to allow immunoprecipitation of the receptors.
7. During incubation, remove 100 μL of 50% slurry of agarose precoupled to an anti-Flag antibody. Place beads into a tube and wash twice with 1 mL of solubilization buffer. Resuspend the beads in 200 μL of solubilization buffer and aliquot 50 μL of the beads into four 1.5-mL tubes. Add to the tube 900 μL of solubilization buffer.
8. After the immunoprecipitation, wash the beads three times with HNTG (1 mL each).
9. After the last wash, transfer 100 μL of beads resuspended in HNTG into a fresh tube using a trimmed pipet tip (this is 10% of the amount of receptor subjected to immunoprecipitation). Pellet the beads by short centrifugation, aspirate the buffer and leave approx 40 μL of buffer on top of the beads. Add to the beads 8 μL of 6X protein gel sample buffer. This is the first IP sample.
10. Centrifuge the remaining beads for a short spin, aspirate the medium, and elute the immunoprecipitated EGFR from the beads by adding 50 μL of IP elution buffer, mixing the tube, and incubating at 95°C for 5 min (make a small hole in the lid of each tube using a needle). Thereafter, centrifuge the tube briefly at 14,000 rpm.

11. Transfer the supernatant to a tube containing agarose beads conjugated to anti-Flag antibodies. Incubate the tube, while gently rotating, for 1 h at 4°C, to allow immunoprecipitation of the receptors that were conjugated with Flag-ubiquitin.

12. At the time of the incubation, remove 100 μL of a 50% slurry of anti-HA precoupled agarose beads and repeat **step 7**.

13. After the Flag-immunoprecipitation, repeat **steps 8–10** ending with the second IP sample (10% of the anti-Flag IP) and the eluant from the Flag immunoprecipitation.

14. Transfer the supernatant to the tube containing the anti-HA conjugated beads and perform immunoprecipitation for 1 h.

15. Wash the IP as above, and at the last wash leave approx 40 μL of buffer on top of the beads. Add to the beads 8 μL of 6X protein gel sample buffer, and heat the sample as above. This is the third IP sample.

16. At the end of the immunoprecipitation steps there will be three protein samples from each plate: first IP (10% of the anti-EGFR IP); second IP (10% of the anti-FLAG IP); and third IP (the whole anti-HA IP). Make a small hole in the lid of each tube using a needle. Heat the samples at 95°C for 5 min; centrifuge briefly.

### 3.2.3. Western Blot Analysis

1. Separate the proteins by SDS-PAGE using 6.5% acrylamide gels (*see* **Note 7**), and electrophoretically transfer the proteins onto a nitrocellulose membrane (as detailed under **Subheading 3.1.4., steps 8 and 9**).

2. Block the membrane by incubating in a blocking solution for 30 min at room temperature (with shaking).

3. Remove the blocking solution and wash the membrane three times with TBST buffer. Incubate the membrane with an antibody directed to the EGFR overnight at 4°C with constant shaking. Thereafter, follow the protocol under **Subheading 3.1.4., steps 12 and 13**.

4. In order to confirm that the immunoprecipitated receptors in the third IP are indeed conjugated with both HA- and Flag-ubiquitins, antibodies on the membranes are stripped by incubating the membrane in a stripping buffer for 30 min at 50°C. Extensively wash the blots with TBST, block the membrane as described previously and reprobe the blot with an anti-Flag antibody, and then with an anti-HA antibody.

## 4. Notes

1. Because reticulocyte lysate contains ubiquitin, one potential concern is that the endogenous wild-type ubiquitin may be incorporated in mixed chains with the ectopic, peptide-tagged ubiquitin, which could complicate interpretation of the results. However, the level of ectopic ubiquitin used in this assay is considerably in excess and therefore competes out the endogenous protein in the reaction. This can be controlled by examining receptor ubiquitination in the absence of ectopic ubiquitin. An alternative approach is to use an ubiquitin-free fraction of reticulocyte lysates (fraction II) *(9)*. A more stringent approach is to employ individual components, i.e., recombinant E1, specific E2s, recombinant c-Cbl, and ubiquitin. In this case, the reaction generates predominantly a monoubiquitylated receptor,

regardless of whether wild-type or KO ubiquitin is used, which indicates that further factors are probably involved in causing multiubiquitination *(8)*.

2. The in vitro system is useful to study other aspects of EGFR ubiquitination. For instance, the assay can be employed to determine whether a given protein is an EGFR-directed E3 ligase. Here, c-Cbl is simply replaced with the putative ligase. Also, the method can be used to clarify the regions in the receptor directly involved in its ubiquitination. To achieve this, receptors are mutated as desired, produced in highly expressing cells such as human kidney cells (HEK-293T) or Chinese hamster ovary (CHO) cells, and subjected to the in vitro reaction.

3. All prepared proteins (e.g., GST-c-Cbl and HA-tagged ubiquitin) should be kept frozen at −70°C. These proteins are sensitive to repeated cycles of freezing and thawing, and thus it is better to freeze them in small aliquots that will be thawed prior to each experiment.

4. The 4KR ubiquitin mutant used in receptor ubiquitination assays is mutated at four lysine residues, which are known to serve as elongation sites. It does, however, retain three additional lysines at positions 6, 27, and 33. These residues were not modified for the in vivo assay because a lysine-less mutant is generally observed to be poorly expressed. In contrast, the in vitro ubiquitination assay enables the use of a lysine-less (KO) mutant since its expression and efficacy in the reaction cascade is unaffected compared to the wild-type protein.

5. A number of controls may be used to validate the specificity of the in vitro reaction. First, the assay can be performed in the absence of ATP or in the presence of an inactive form of Cbl in which the E2 or substrate recognition sites are defective *(6,10)*. Under such conditions, very little or no ubiquitination should be observable. Second, the efficacy of KO ubiquitin as a chain terminator can be checked using substrates known to undergo polyubiquitination but not multiubiquitination (e.g., β-catenin, IκB, or p105 of NF-κB).

6. The intensity of the signal can affect the analysis of such experiments. If too weak, the signal can be improved by calibrating increasing amounts of ubiquitin in the reaction mixture. Alternatively, a more sensitive approach is to radiolabel the recombinant ubiquitins with [125]I prior to the reaction. In this case, the polyacrylamide gel is dried after electrophoresis and exposed directly to x-ray film.

7. When running gels to be blotted for ubiquitination, it is advisable to use 10-well combs rather than the 15-well combs. The wider lanes better resolve high molecular weight smears of ubiquitylated proteins.

8. A simpler version of the sequential triple IP method is to perform only a double IP, for example, for the receptor and one of the ubiquitin tags, and blot for the second tag. Thus, a receptor is isolated that is conjugated to the 4KR ubiquitin mutant, and subsequent detection using an antibody directed to the other tag is indicative of multiubiquitination. This approach is simpler and yields a stronger signal, but on the other hand, the chances for increased background are also higher.

9. Alternative tags, other than HA and Flag (e.g., His or Myc), can be used. However, there are two points to take into consideration when choosing a tag. First, the tag should be applicable both for immunoprecipitation and for immunoblotting. Sec-

ond, a peptide tag should retain its ability to be immunoprecipitated even after heating at 95°C. GST tag, for example, does not bind glutathione after such treatment and therefore cannot be pulled down.

10. Other cell lines can be used in ubiquitination experiments, but it is better to use highly expressing cells since proteins are analyzed after three rounds of IP. In addition, it is recommended to use cells that express low levels of endogenous EGFR, such as CHO or HEK-293T, in order to prevent saturation of the beads with receptors from cells that were not transfected with the peptide-tagged ubiquitin.

## Acknowledgments

We thank members of our group for useful insights and feedback. Our laboratory is supported by research grants from The Prostate Cancer Foundation, the National Cancer Institute (grant CA72981), and the Israel Science Foundation. Y. Yarden is the incumbent of the Harold and Zelda Goldenberg Professorial Chair. The Willner Family Center for Vascular Biology is a long-term supporter of our laboratory.

## References

1. Weissman, A. M. (2001) Themes and variations on ubiquitination. *Nat. Rev. Mol. Cell. Biol.* **2(3)**, 169–178.

2. Pickart, C. M. (2000) Ubiquitin in chains. *Trends Biochem. Sci.* **25(11)**, 544–548.

3. Thrower, J. S., Hoffman, L., Rechsteiner, M. and Pickart, C. M. (2000) Recognition of the polyubiquitin proteolytic signal. *EMBO J.* **19(1)**, 94–102.

4. Hicke, L. (2001) Protein regulation by monoubiquitin. *Nat. Rev. Mol. Cell. Biol.* **2(3)**, 195–201.

5. Levkowitz, G., Waterman, H., Zamir, E., et al. (1998) c-Cbl/Sli-1 regulates endocytic sorting and ubiquitination of the epidermal growth factor receptor. *Genes Dev.* **12(23)**, 3663–3674.

6. Levkowitz, G., Waterman, H., Ettenberg, S. A., et al. (1999) Ubiquitin ligase activity and tyrosine phosphorylation underlie suppression of growth factor signaling by c-Cbl/Sli-1. *Mol. Cell* **4**, 1029–1040.

7. Haglund, K., Sigismund, S., Polo, S., Szymkiewicz, I., Di Fiore, P. P. and Dikic, I. (2003). Multiple monoubiquitination of RTKs is sufficient fr their endocytosis and degradation. *Nat. Cell. Biol.* **5(5)**, 461–466.

8. Mosesson, Y., Shtiegman, K., Katz, M., et al. (2003) Endocytosis of receptor tyrosine kinases is driven by monoubiquitination, not polyubiquitination. *J. Biol. Chem.* **278(24)**, 21,323–21,326.

9. Hershko, A., Heller, H., Elias, S. and Ciechanover, A. (1983) Components of ubiquitin-protein ligase system. Resolution, affinity purification, and role in protein breakdown. *J. Biol. Chem.* **258(13)**, 8206–8214.

10. Yokouchi, M., Kondo, T., Houghton, A., et al. (1999) Ligand-induced ubiquitination of the epidermal growth factor receptor involves the interaction of the c-Cbl RING finger and UbcH7. *J. Biol. Chem.* **274**, 31,707–31,712.

# 9

## Assays to Monitor Degradation of the EGF Receptor

### Mirko H. H. Schmidt and Ivan Dikic

#### Summary

Recently the life cycles of receptor tyrosine kinases (RTKs) have become a focus of signal transduction research. Ligand-induced ubiquitination of RTKs followed by their internalization and degradation has, in particular, been extensively studied. This chapter describes the basic methods used to measure ubiquitination and degradation of RTKs using the example of the epidermal growth factor receptor (EGFR). Common sources for endogenous and recombinant EGFR as well as cell lines used to conduct receptor downregulation assays are described. Monitoring of ubiquitination and degradation of the EGFR subsequent to stimulation with the receptor ligand EGF is described. Finally, protocols to quantitatively measure degradation of the EGFR by pulse chase experiments or using radiolabeled ligands such as $^{125}$I-EGF are presented.

**Key Words:** Casitas B-lineage lymphoma (Cbl); degradation; downregulation; endocytosis; epidermal growth factor receptor (EGFR); internalization; monoubiquitination; polyubiquitination; receptor tyrosine kinase (RTK); ubiquitin; ubiquitination.

## 1. Introduction

Receptor tyrosine kinases (RTKs) are cell membrane-spanning proteins that transmit signals via the cellular membrane by binding to specific stimulators on the outside of the cell *(1)*. Association with such a ligand often causes oligomerization of the receptor that initiates an intracellular signaling cascade, which in turn affects cell proliferation, cellular adhesion or apoptosis *(1)*. In order to regulate the lifetime of this system, activated RTKs also initiate a negative-feedback loop that eventually leads to the removal of the RTK complex from the cell membrane by endocytosis *(2)*. Recent studies have shown that this process is mediated by ubiquitin ligases such as the members of the Cbl protein family, which are recruited to activated RTKs (i.e., the EGFR) and are activated by phosphorylation during this process *(3,4)*. This leads to the translocation of intracellular C-Cbl-interacting protein of 85 kDa (CIN85)-endophilin

From: *Methods in Molecular Biology, vol. 327: Epidermal Growth Factor: Methods and Protocols*
Edited by: T. B. Patel and P. J. Bertics © Humana Press Inc., Totowa, NJ

complexes to Cbl, and by the action of endophilins it causes a negative curvature of the plasma membrane, which initiates endocytosis *(5,6)*. Cbl-mediated multiple monoubiquitination of RTKs that occurs in parallel directs the receptor containing vesicle to a degradation route and thereby leads to the termination of the RTK signal *(7,8)*.

In this chapter we will present methods that can be used to display and measure the processes described above. Expression and immunoprecipitation studies will show how the EGFR becomes associated with Cbl proteins subsequent to stimulation with EGF and how this leads to the covalent attachment of ubiquitin moieties to the receptor. In parallel, the amount of EGFR as detected in Western blots decreases. We will present a protocol that describes how this decrease can be quantified by a pulse chase assay. Finally, we will show how the amount of receptor physically present at the cellular surface can be measured by radiolabeled $^{125}$I-EGF using a gamma counter and how this value decreases over time when the EGFR is exposed to its ligand EGF. We describe three major variations of how to monitor ligand-induced downregulation of RTKs using as an example EGFR. (For additional approaches and methodologies to assess receptor downregulation in general, the reader is referred to Chapters 8 and 10.)

## 2. Materials

### 2.1. Immunoprecipitation and Immunoblotting

#### 2.1.1. Materials for the Expression of Recombinant Proteins in Mammalian Cells

1. Dulbecco's modified Eagle's medium (DMEM) + penicillin/streptomycin + 10% fetal calf serum (FCS).
2. Serum-free DMEM.
3. DMEM + 2X penicillin/streptomycin + 20% FCS.
4. Mammalian expression plasmids encoding for EGFR, Cbl, and Flag-tagged ubiquitin.
5. Transfection reagents such as lipofectamine (Invitrogen; *see* **Note 1**).
6. 100 mg/mL human recombinant EGF (BD, Sigma) stock solution.
7. Liquid nitrogen.

#### 2.1.2. Immunoprecipitation

1. Lysis buffer: 50 m$M$ hydroxyethyl piperazine ethane sulfonate (HEPES) (pH 7.5), 150 m$M$ NaCl, 1 m$M$ ethylenediaminetetraacetic acid (EDTA), 1 m$M$ ethylene glycol tetraacetic acid (EGTA), 10% glycerol, 1% Triton X-100, 25 m$M$ NaF, 10 μ$M$ ZnCl$_2$. Store at 4°C. Prior to use add 1 m$M$ sodium orthovanadate, 1 m$M$ PMSF, 10 μg/mL aprotinin and 2 μg/mL leupeptin (*see* **Note 2**).
2. Anti-EGFR, anti-Cbl, and anti-Flag (M2, Sigma) antibodies (*see* **Note 3**).
3. Protein A-Agarose conjugate (Roche; *see* **Note 4**).

4. Sodium dodecyl sulfate-polyacrylamide gel electrophoresis (SDS-PAGE) sample buffer (*see* **Note 5**).

### 2.1.3. Immunoblotting (see **Note 6**)

1. SDS-PAGE and blotting system (BioRad, Invitrogen).
2. Prestained protein markers (BioRad, Invitrogen).
3. Nitrocellulose membrane 0.4 μm (Osmonics)
4. Anti-EGFR, anti-Cbl, anti-Flag (M2, Sigma), and anti-phosphotyrosine antibodies.
5. Horseradish peroxidase (HRP)-conjugated secondary antibodies and HRP-conjugated protein A.
6. Tris-buffered saline (TBS): 20 m$M$ Tris-HCl (pH 7.5), 150 m$M$ NaCl.
7. TBST: 0.5% Triton X-100 in TBS.
8. Blocking solution 1: 5% bovine serum albumin (BSA) in TBS.
9. Blocking solution 2: 5% milk powder in TBS.
10. Chemiluminescence detection system.

## 2.2. EGFR Pulse Chase Assay

1. *See* **Subheading 2.1.**
2. Promix (Amersham-Pharmacia).
3. Gel drying system (BioRad).
4. Phosphoimager system (BioRad).
5. Cycloheximide (optional).

## 2.3. $^{125}$I-EGF Receptor Downregulation Assay

1. F12K medium (Kaighn's modification) + penicillin/streptomycin + 10% FCS.
2. Serum-free F12K.
3. Mammalian expression plasmids encoding for EGFR and Cbl.
4. Transfection reagents such as lipofectamine (Invitrogen; *see* **Note 1**).
5. Internalization medium: F12K + 0.1% BSA + 10 m$M$ HEPES (pH 7.5).
6. Stripping solution: phosphate-buffered saline (PBS) + 0.1% BSA (pH 3.4 with acetic acid).
7. Cell lysis solution: 1 $M$ NaOH (*see* **Note 7**).
8. 100 mg/mL human recombinant EGF (BD, Sigma) stock solution.
9. 5 μCi $^{125}$I-EGF human recombinant EGF (Amersham-Pharmacia; *see* **Note 8**).
10. Gamma counter (1470 Wizard, Perkin Elmer; *see* **Note 9**).

## 3. Methods

### 3.1. EGFR Downregulation Measured by Immunoprecipitation and Immunoblotting (see Note 10)

#### 3.1.1. Expression in Mammalian Cells (HEK293) and Incubation With EGF

1. Human embryonic kidney (HEK)293 cells (*see* **Note 11**) are cultured under standard conditions (37°C, 5% $CO_2$, 95% humidity) in DMEM (10% FCS plus antibiotics) and are harvested 3 d prior to the experiment.
2. 200,000 to 500,000 cells per well are seeded in a six-well plate.

3. The next day, at 80% confluence, cells are transfected with 1 μg of EGFR encoding plasmid (pRK5-EGFR), 0.5 μg of Cbl encoding plasmid (pRK5-Cbl), and 1 μg of Flag-tagged ubiquitin encoding plasmid (pcDNA3-Ub-Flag) (*see* **Note 12**) using 5 μL of lipofectamine (Invitrogen) per well in 1.2 mL of serum-free medium according to the manufacturer's guidelines (*see* **Note 1**).
4. Five hours later 1.2 mL of medium containing the double amount of serum and antibiotics is added (*see* **Note 13**).
5. The next day medium is very carefully exchanged to 1 mL of serum-free medium.
6. Twenty hours later EGF-stimulation experiments are performed (*see* **Note 14**).
7. 1 mL of serum-free medium containing 100 ng/mL of human recombinant EGF is added and cells are incubated at 37°C for individual time points, usually 0, 2, 5, 10, 30, and 120 min. The well marking time point zero receives 1 mL of serum-free medium only (*see* **Note 15**).
8. After the incubation medium is removed, cells are frozen in liquid nitrogen. Cells can be stored at –80°C until immunoprecipitation is performed.

### 3.1.2. Cell Lysis and Immunoprecipitation

1. Frozen cells on a six-well plate are thawed on ice (*see* **Note 16**). Then 1 mL of lysis buffer is added and the suspension is incubated for 10 min on ice.
2. Cells are scraped off with a cell lifter and transferred to a 1.5-mL tube.
3. Tubes are centrifuged for 15 min at 4°C and 16,200$g$ in order to remove cellular debris.
4. Supernatant is transferred to a fresh tube and 50 μL of the solution is mixed with SDS-PAGE sample buffer (called total cell lysate [TCL]).
5. The rest is incubated with an anti-EGFR antibody known to have high specificity and binding efficiency (i.e., 0.3 μL MAb 108) and rotated at 4°C for 30 min. The amount of other antibodies needs to be titrated.
6. Subsequently, 30 μL of protein A-agarose conjugate (Roche; *see* **Note 4**) are added and the suspension is incubated for 2 h up to overnight at 4°C.
7. Agarose beads are collected by centrifugation for 2 min at 4°C and 16,200$g$.
8. Supernatant is removed, beads are washed with 1 mL of lysis buffer, and recollected by centrifugation.
9. After the third washing step, supernatant is removed and beads are centrifuged for 30 s at 4°C and 16,200$g$. Excessive liquid is removed by vacuum suction.
10. Agarose beads are suspended in 30 μL of SDS-PAGE sample buffer (*see* **Note 5**) and heated at 95°C for 5 min.
11. Subsequently, samples are transferred on ice and may be stored at –80°C until further analysis (*see* **Note 17**).

### 3.1.3. Immunoblotting (see **Note 6**)

1. SDS-PAGE should be performed according to standard procedures. In brief, prepare gels with 4% stacking gel and 7% separation gel in a Tris-glycine system or use precasted 4–12% gradient gels (Invitrogen).
2. Thaw samples on ice and boil at 95°C for 5 min. Then cool on ice and precipitate agarose beads by centrifugation for 5 min at 4°C and 16,200$g$.

3. Load 10 µL of sample solution per well and flank samples by prestained markers (*see* **Note 18**).

4. Run Tris-glycine gels for approx 1 h at 0.02 A per gel, Invitrogen gels according to the manufacturer's guidelines.

5. Transfer proteins from the gel onto nitrocellulose membrane (*see* **Note 19**) using a standard blotting procedure.

6. Block membrane for >30 min with blocking solution 1.

7. Incubate membrane for 1 h with anti-EGFR antibody or anti-Cbl antibody in blocking solution 1 to detect EGFR and Cbl levels in immunoprecipiates and TCL. Additionally, use anti-Flag antibody (M2, Sigma) to detect attachment of Flag-ubiquitin to EGFR molecules in immunoprecipitates and TCL. Additionally, detect EGFR phosphorylation by using any phosphotyrosine-specific antibody (i.e., PY99, Santa Cruz Biotechnology).

8. Remove the antibody solution and wash membranes four times for 5 min with TBST.

9. Incubate membranes for 1 h with a 1:6000 dilution of secondary antibody or a 1:7000 dilution of HRP-conjugated protein A in blocking solution 2.

10. Remove antibody solution and wash membranes two times for 5 min with TBST and three times for 10 min in TBS.

11. Develop the blots using any chemiluminescence system (*see* **Note 20**).

### *3.2. Monitoring EGFR Levels by Pulse Chase (see Note 21)*

1. HEK293 cells are transfected with EGFR and Cbl expression plasmid (*see* **Subheading 3.1.1.**).

2. The next day cells are incubated at 37°C in DMEM depleted of cysteine and methionine (Sigma) for 1 h.

3. Medium is removed and cells are labeled with Promix (50–100 µCi/mL) in 1 mL of standard serum-free medium containing cysteine and methionine for 2–5 h.

4. EGF stimulation is performed with 50 ng/mL EGF as described above (*see* **Note 22**).

5. Medium is removed, cells are washed two times in ice-cold serum-free medium. Cells are lysed, EGFR is immunoprecipitated (*see* **Subheading 3.1.2.**) and proteins are resolved by SDS-PAGE (*see* **Subheading 3.1.3.**).

6. The gel is dried for 2 h at 85°C using a gel-drying system (BioRad).

7. The $^{35}$S signal of precipitated EGFR is measured and quantified using a phosphoimager system (BioRad) (*see* **Note 23**).

### *3.3. Quantification of Membrane Standing EGFR by $^{125}$I-EGF Labeling (see Note 8)*

1. Chinese hamster ovary (CHO)-K1 cells are cultured under standard conditions (37°C, 5% $CO_2$, 95% humidity) in F12K medium (Kaighn's modification), including 10% FCS plus antibiotics and are harvested 3 d prior to the experiment.

2. 1,000,000 cells are seeded in a 10-cm cell culture dish.

3. The next day at 80% confluence cells are transfected with 2 µg of EGFR expressing plasmid (pRK5-EGFR) and 1.5 µg of Cbl expressing plasmid (pRK5-Cbl)

(*see* **Note 12**) using 15 μL of lipofectamine (Invitrogen) per well in 6 mL of serum-free medium according to the manufacturer's guidelines (*see* **Note 24**).

4. Five hours later, medium is removed and changed to standard incubation medium.
5. The next day, cells are split into the wells of a 12-well plate.
6. At d 3 cells are washed two times with serum-free medium.
7. Internalization of the EGFR is initiated with 50 ng/mL of EGF for various time points between 0 and 60 min. Each time point is measured in triplicate.
8. Plates are transferred on ice, and each well is incubated with ice-cold stripping solution for 5 min to remove surface-bound EGF.
9. Plates are washed two times with ice-cold internalization medium.
10. Cells are incubated with 500 μL of internalization medium contain approx 100,000 counts of $^{125}$I-EGF for 1 h.
11. Supernatant is removed, and cells are washed two times with ice-cold internalization medium.
12. Cells are lysed with 500 μL of 1 *M* NaOH solution (*see* **Note 7**) for 15 min.
13. Lysates are measured in a γ counter (1470 Wizard, Perkin Elmer; *see* **Note 9**).

## 4. Notes

1. Any other transfection reagent may be used. Calcium-phosphate-based methods work fine for HEK293 cells. In brief, 1–2 million HEK293 cells are seeded per 10-cm dish and transfection is performed the next day. A mixture of 437 μL of TE buffer (10 m*M* Tris, pH 7.9, and 0.1 m*M* EDTA) containing a maximum of 20–30 μg DNA is combined with 63 μL of 2 *M* CaCl$_2$ solution. Solution is slowly added to 500 μL of 2X HBS solution (50 m*M* HEPES [pH 7.12], 280 m*M* NaCl, 10 m*M* KCl, and 1.5 m*M* Na$_2$HPO$_4$) along a 1-mL pipet under permanent bubbling in order to form a fine DNA precipitate. Please note that the pH of HBS depends on the cell line to be transfected. Solution is added to cells under permanent shaking of the cell culture dish and incubated in the cell culture incubator for 4–5 h. Subsequently, medium is replaced by fresh medium, and cells may be harvested after 24 to 48 h.
2. Phosphatase and proteinase inhibitors should be stored in separate containers at −20°C until used. PMSF is unstable in water and toxic but can be replaced by AEBSF (Sigma) or Pefabloc SC (Fluka).
3. Anti-Flag M5 antibody may also be used, but is more sensitive than M2 and can result in a strong background staining of the blot.
4. Any other protein A–agarose is fine. Some antibodies do not bind to protein A. In this case protein G–agarose should be used. Chicken antibodies are difficult overall in immunoprecipitation experiments. A protein A/G combination can be used, but usually with a low pull-down efficiency.
5. Any self-made or commercial Laemmli buffer is fine. However, one should take care that the agarose beads are resolved in a 2X concentrated SDS-PAGE sample buffer.
6. Other blotting methods should be fine as well. However, we achieved highest sensitivity through the combination of Tris-glycine or Invitrogen gels for SDS-PAGE, nitrocellulose membrane for blotting, HRP-conjugated secondary antibodies for immunodetection, and a chemiluminescence system for visualization.

7. Any other lysis buffer may be used.

8. Special caution is necessary while working with γ radiation; please contact your radiation safety department for further information. In brief, a 3-mm lead foil is adequate to block γ radiation. However, evaporation of radioactive material is critical since it might be incorporated into the thyroid. Therefore, experiments should be done in a hood.

9. Liquid scintillation counters able to detect $^{125}$I can be used.

10. In this assay it is demonstrated how immunoprecipitated EGFR becomes ubiquitinated and degraded in response to EGF stimulation. However, the effect is also measurable in the TCL fractions. Loading approximately 5 µL of TCL and blotting for EGFR while coexpressing pcDNA3-Ub-Flag leads to an EGFR band that becomes "smeary" after stimulation with EGF. This is due to the attachment of ubiquitin moieties and secondary modifications of the receptor but not to polyubiquitination, because co-expression of a pcDNA3-Ub-mut-Flag vector that encodes an ubiquitin molecule with three mutations thus unable to form ubiquitin chains causes the same "smeary" pattern. Eventually, EGFR band and smear start to disappear subsequent to stimulation with EGF.

11. Any cell line can be used to perform this experiments, but HEK293 have been shown to be easily transfected and handled. However, in order to avoid transfection of recombinant EGFR, cell lines can be used that express high amounts of endogenous EGFR, such as HeLa cervix carcinoma cell line, SW480 colon carcinoma cell line, as well as U87 and LNZ308 glioma cell lines. These cell lines are more difficult to transfect (Invitrogen's LP2000 or Roche's Fugene 6 may be used), and they are of tumorogenic origin, which represents a problem for the subsequent data analysis. Signaling pathways in these cell lines might be altered in a tumor-type-specific manner.

12. DNA amount depends strongly on the expression plasmid used and must be tested up front.

13. Transfection medium can also be removed and be replaced by standard cultivation medium containing FCS. However, the proposed procedure is preferred because HEK293 cells do easily detach during medium exchange. This loss of cells leads to variations in the amount of transfected cells and makes it more difficult to reach a stable expression level of recombinant proteins.

14. At this stage HEK293 cells are barely attached to the cultivation dish. Cells must be handled very carefully because every bump may lead to the complete detachment of cells, which terminates the experiment.

15. The no-EGF control also receives serum since the mechanical force caused by pipetting can induce a weak activation of transfected EGFR.

16. Cells will look "wet" when thawed.

17. Freezing and storing time should be minimal since phosphates tend to fall off their proteins, which is a problem if the phosphorylation status of a protein is to be checked.

18. The markers will help you later on if you intend to cut your membrane into several pieces to detect multiple proteins on a single blot.

19. Polyvinylidine fluoride (PVDF) membrane will also work but cannot effectively be stripped for further use.
20. Afterwards blot may be stripped of antibody for 10 min in 0.2 $M$ NaOH solution to be reused.
21. Special caution is necessary while working with $^{35}$S; please contact your radiation safety department for further information. Evaporation of radioactive material is critical, and all experiments should be done in a hood.
22. Effects can be intensified by addition of 60 µg/mL cycloheximide to the cell culture medium for an additional 1 h before and during the EGF-stimulation experiment (CHX chase).
23. If you do not have a gel dryer and phosphoimager system, you can also perform a Western blot after the SDS-PAGE and detect the $^{35}$S signal by an x-ray film.
24. The cell density is very critical. If CHO cells are too dense, they will not effectively express any transfected plasmid. If cells are not dense enough, they will die from the lipofectamine. If you are not experienced, you should test your transfection protocol before running a big experiment.

## References

1. Ullrich, A. and Schlessinger, J. (1990) Signal transduction by receptors with tyrosine kinase activity. *Cell* **61,** 203–212.
2. Dikic, I. and Giordano, S. (2003) Negative receptor signalling. *Curr. Opin. Cell. Biol.* **15,** 128–135.
3. Dikic, I., Szymkiewicz, I., and Soubeyran, P. (2003) Cbl signaling networks in the regulation of cell function. *Cell. Mol. Life Sci.* **60,** 1805–1827.
4. Thien, C. B. and Langdon, W. Y. (2001) Cbl: many adaptations to regulate protein tyrosine kinases. *Nat. Rev. Mol. Cell. Biol.* **2,** 294–307.
5. Soubeyran, P., Kowanetz, K., Szymkiewicz, I., Langdon, W. Y., and Dikic, I. (2002) Cbl-CIN85-endophilin complex mediates ligand-induced downregulation of EGF receptors. *Nature* **416,** 183–187.
6. Petrelli, A., Gilestro, G. F., Lanzardo, S., Comoglio, P. M., Migone, N., and Giordano, S. (2002) The endophilin-CIN85-Cbl complex mediates ligand-dependent downregulation of c-Met. *Nature* **416,** 187–190.
7. Haglund, K., Sigismund, S., Polo, S., Szymkiewicz, I., Di Fiore, P. P., and Dikic, I. (2003) Multiple monoubiquitination of RTKs is sufficient for their endocytosis and degradation. *Nat. Cell. Biol.* **5,** 461–466.
8. Haglund, K., Di Fiore, P. P., and Dikic, I. (2003) Distinct monoubiquitin signals in receptor endocytosis. *Trends. Biochem. Sci.* **28,** 598–603.
9. Mosesson, Y., Shtiegman, K., Katz, M., et al. (2003) Endocytosis of receptor tyrosine kinases is driven by monoubiquitylation, not polyubiquitylation. *J. Biol. Chem.* **278,** 21,323–21,326.

# 10

## Assessment of Degradation and Ubiquitination of CXCR4, a GPCR Regulated by EGFR Family Members

### Adriano Marchese

### Summary

The ErbB2/HER2 receptors are aberrantly expressed in certain mammary epithelial cancers. A recent study has shown that in a subset of these breast cancers, the HER2 receptors contribute to increased cell surface levels of the chemokine receptor CXCR4, which in turn results in increased metastasis of the breast cancers. Therefore, information concerning the mechanisms regulating CXCR4 receptors levels is essential to our understanding of its role in cancer cell metastasis. CXCR4 is a member of the G protein-coupled receptor (GPCR) family, and herein we describe methods to monitor the ubiquitination, degradation, and downregulation of CXCR4.

**Key Words:** ErbB2/HER2; CXCR4; chemokine; ubiquitination; endocytosis; SDF-1; downregulation; G protein-coupled receptor.

## 1. Introduction

Multiple factors contribute to cancer metastasis, but detailed cellular and mechanistic insight into this process is lacking. Recently, a subset of chemokines and their cognate receptors has been implicated in breast cancer metastasis (*1*). Chemokines belong to a group of low molecular weight, cytokine-like proteins that act through G protein-coupled receptors (GPCRs) to mediate leukocyte chemotaxis (*2*). Recent developments have revealed that the chemokine receptor CXCR4, whose cognate ligand is stromal-cell-derived factor $1\alpha$ (SDF-$1\alpha$; CXCL12), is overexpressed in a subset of breast cancers, which appears to contribute to the metastatic potential of breast cancer cells to organs that express CXCL12 (*1*). In addition, CXCR4 expression appears to be increased in several other types of cancer including lung (*3,4*), prostate (*5*), and pancreatic cancer (*6*). However, the mechanisms contributing to increased expression of CXCR4 remain largely unknown.

From: *Methods in Molecular Biology, vol. 327: Epidermal Growth Factor: Methods and Protocols*
Edited by: T. B. Patel and P. J. Bertics © Humana Press Inc., Totowa, NJ

CXCR4 expression levels are regulated by multiple pathways, including at the level of transcription *(7)*, translation *(8)*, and posttranslationally *(9)*. Intriguingly, a recent study has suggested that aberrant regulation of CXCR4 protein levels may contribute to metastasis of some cancers in which HER2 is overexpressed *(8)*. HER2 is a receptor tyrosine kinase belonging to the epidermal growth factor receptor family *(10)*. It is activated by homodimerization or by growth factors upon heterodimerization with other members of the ErbB2 family *(10)*. Aberrant HER2 expression is observed in approx 30% of breast cancers and is an important therapeutic target in breast cancer therapy *(11)*. Herceptin, an anti-HER2 monoclonal antibody, has proven to be effective therapy in breast cancers in which HER2 is overexpressed. A recent study has linked HER2 overexpression to CXCR4 overexpression in a subset of breast cancers *(8)*. It appears that aberrant HER2 expression impairs the ability of CXCR4 to become ubiquitinated and thus degraded, thereby contributing to higher than normal cell surface levels of CXCR4. The increase in CXCR4 levels contributes towards increased metastasis of the cancer cells. This link between overexpression of HER2 and increase in cell surface CXCR4 underscores the need to study the regulation of CXCR4 turnover with a view of designing strategies to decrease CXCR4 activity or amounts in breast cancers.

CXCR4 regulation is complex and involves the sequential and concerted action of several distinct proteins. Upon agonist stimulation, CXCR4 undergoes rapid internalization through clathrin-coated pits in a G protein-coupled receptor kinase (GRK)- and arrestin-dependent manner *(12)*. Also, CXCR4 is regulated by ubiquitin modification upon agonist activation (*see* **Fig. 1**). CXCR4 undergoes agonist-dependent monoubiquitination, which serves as an endosomal sorting signal targeting the receptor to lysosomes for degradation *(9)*. Mechanistic insight into this process has revealed that the Nedd4-like HECT domain E3 ubiquitin ligase AIP4 mediates agonist-dependent ubiquitination of CXCR4 at the plasma membrane, revealing that the ubiquitination status of the receptor correlates with its ability to be degraded *(13)*. Recently a new study revealed that HER2 directly regulates CXCR4 receptor levels by blocking the ability of the receptor to become ubiquitinated and thus degraded, thereby contributing to increased CXCR4 levels on the surface of the plasma membrane, which then leads to greater cancer metastasis *(8)*.

Here we describe relatively straightforward methods that we have used to detect CXCR4 degradation and ubiquitination in a rapid and highly accurate manner. These protocols can be applied to any number of receptors either overexpressed or endogenous from several different cell lines.

## 2. Materials
1. HeLa cells (American Tissue Type Collection) and human embryonic kidney cells (Microbix, Toronto, ON).

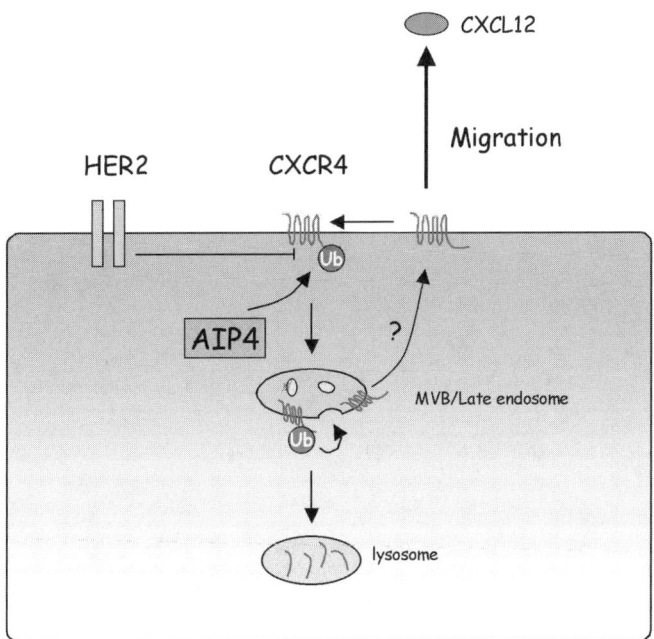

Fig. 1. Overexpression of HER2 contributes to increased cell surface levels of CXCR4 leading to increased cell migration. Upon agonist (CXCL12) binding, CXCR4 is internalized via clathrin-coated pits through an arrestin- and GRK-dependent pathway *(12)*. In addition, upon agonist binding CXCR4 is rapidly monoubiquitinated by the E3 ubiquitin ligase AIP4 *(13)*. The ubiquitin moiety (Ub) serves as a sorting signal to enter the multivesicular body (MVB) for subsequent degradation of the receptor in lysosomes. Overexpression of HER2 attenuates the ubiquitination of CXCR4 by an unknown mechanism and subsequent targeting to lysosomes, leading to increased levels of CXCR4 *(8)*. Mistargeting to lysosomes could potentially lead to recycling of CXCR4 back to the cell surface. The increased surface expression of CXCR4 makes the cells more responsive to CXCL12 stimulation and can result in increased metastasis.

2. DNA expression constructs: HA-tagged CXCR4 in pcDNA3; 3X FLAG-tagged ubiquitin in pCMV-10.
3. Antibodies: anti-CXCR4 (2B11); anti-HA (Covance, Berkeley, CA); anti-FLAG M2 (Sigma, St. Louis, MO); anti-β-tubulin (Accurate Chemical and Scientific Corporation, Westbury, NY); anti-mouse IgG conjugated with horseradish peroxidase (Vector Laboratories, Burlingame, CA); anti-rat IgG conjugated with horseradish peroxidase (Jackson Lab).
4. Phosphate-buffered saline (PBS): 137 m$M$ NaCl, 2.7 m$M$ KCl, 10.22 m$M$ $Na_2HPO_4$, 1.7 m$M$ $KH_2PO_4$ (pH 7.4).
5. 2X Sample buffer for electrophoresis: 0.0375 $M$ Tris-HCl (pH 6.5), 8% sodium dodecyl sulfate (SDS), 10% glycerol, 5% β-mercaptoethanol, 0.003% bromophenol blue.

6. 10% SDS–polyacrylamide (PAGE) gels
7. Tris-buffered saline (50 m$M$ Tris-HCl [pH 7.4], 150 m$M$ NaCl) containing 0.05% Tween-20 (TBST). Also, TBST containing 5% (w/v) nonfat dry milk (TBST–5% milk).
8. Lysis buffer: 50 m$M$ Tris-HCl (pH 8.0), 150 m$M$ NaCl, 5 m$M$ ethylenediametetraacetic acid (EDTA), 0.5% sodium deoxycholate (w/v), 1% nonidet P-40 (NP40; v/v), 0.1% SDS (w/v).
9. Protease inhibitors are added fresh each time with the following final concentrations: 10 µg/mL aprotinin, 10 µg/mL leupeptin, 0.2 mg/mL benzamidine, and 1 µg/mL pepstatin-A.
10. NEM is added fresh each time at a final concentration of 20 m$M$.
11. SDS–running buffer (25 m$M$ Tris-HCl [pH 7.4], 0.192 m$M$ glycine, 1% SDS [w/v]).
12. SDS–transfer buffer (25 m$M$ Tris-HCl [pH 7.4], 0.192 m$M$ glycine, 20% methanol).
13. SuperSignal® chemiluminescent substrate (Pierce, Rockford, IL).
14. FuGene6 transfection reagent and protein-A agarose beads (Roche, Indianapolis, IN).
15. Dulbecco's modified Eagle's medium (DMEM) (Mediatech, Herndon, VA) and fetal bovine serum (FBS) (Atlanta Biologicals, Atlanta, GA).
16. Stromal-cell-derived growth factor-1α (SDF, also known as CXCL12, a CXCR4 agonist) (PeproTech, Rocky Hill, NJ) dissolved in DMEM containing 0.5% FBS.

## 3. Methods

### 3.1. Degradation Assay

1. Seed HeLa cells onto 6-cm tissue culture dishes (*see* **Note 1**).
2. After 24 h the cells should be approx 80–90% confluent. Wash cells once with 1 mL of warm DMEM.
3. Incubate cells in 1.5 mL of prewarmed DMEM supplemented with FBS (10% v/v) plus vehicle (DMEM + 0.5% FBS) or 100 n$M$ SDF at 37°C. The time of incubation is typically 3 h (*see* **Note 2**).
4. After stimulation, cells are washed with 2 mL of room-temperature PBS.
5. To lyse cells, add 600 µL of 2X sample buffer directly to cells, scrape, and carefully transfer lysate to a fresh microcentrifuge tube (*see* **Note 3**).
6. Sonicate sample at the lowest setting of a Branson digital sonifier (*see* **Note 4**).
7. Resolve proteins by SDS-PAGE and transfer to nitrocellulose membranes using a standard Western blot protocol.
8. The membrane is then blocked for 30 min in 10 mL of TBST containing 5% (w/v) nonfat dry milk.
9. The nitrocellulose is next incubated with 10 mL of TBST-5% milk containing rat monoclonal anti-CXCR4 antibody at a dilution of 1:1000 overnight at 4°C under gentle agitation.
10. Wash the nitrocellulose three times for 10 min each in TBST.
11. Incubate the nitrocellulose for 30 min with 10 mL of TBST–5% milk containing goat anti-rat IgG conjugated to horseradish peroxidase at a dilution of 1:10,000.
12. Wash the membrane five times for 10 min each in TBST.

13. Overlay the nitrocellulose with 1–2 mL of Supersignal chemiluminescence reagent for approx 5 min, allow the blot to drip-dry, wrap in plastic wrap, and visualize on x-ray film.
14. Parallel blots for tubulin for a loading control should also be analyzed.

## 3.2. Ubiquitination Assay

1. HEK293 cells are maintained in DMEM supplemented with 10% FBS (*see* **Note 5**).
2. Pass cells onto 10-cm dishes for transfection the following day. The cells should be approx 50–80% confluent at the time of transfection.
3. Cells are then transiently transfected using Fugene 6 transfection reagent with 1 µg each of DNA encoding hemagglutinin (HA)-tagged CXCR4 and 3X FLAG-tagged ubiquitin, according to the manufacturer's instructions.
4. Approximately 24 h later, the cells are passaged onto 6-cm dishes.
5. Twenty-four hours later, wash cells once with 2 mL of warm DMEM supplemented with 20 m$M$ hydroxethyl peperazine ethane sulfonate (HEPES), pH 7.4.
6. Incubate cells in the same media in the presence or absence of 100 n$M$ SDF (a CXCR4 agonist) for 30 min (*see* **Note 6**).
7. Place dishes on ice and wash once with 2 mL of ice-cold PBS.
8. Add 1 mL of lysis buffer, scrape cells, and transfer to 1.5-mL microcentrifuge tubes. Incubate tubes on ice for approx 10 min to allow for complete solubilization (*see* **Note 7**).
9. To pellet cellular debris, centrifuge samples at maximum speed in a microcentrifuge for 30 min at 4°C.
10. Carefully transfer 300 µL of supernatant to a fresh microcentrifuge tube. Add 2.5 µL of polyclonal anti-HA antibody and incubate for 1 h while rocking at 4°C.
11. Add 20 µL of protein A–agarose (prepared by diluting 1:1 w/v with lysis buffer) and continue incubation for an additional 1 h while rocking at 4°C.
12. Collect receptor–protein A–agarose complexes by centrifugation in a microcentrifuge at 12,000$g$ for 5 s.
13. Carefully remove supernatant, resuspend beads in 750 µL of lysis buffer, and incubate for 10 min at 4°C while rocking.
14. Repeat **steps 12** and **13**.
15. Collect complexes as in **step 12** and carefully remove the last traces of lysis buffer.
16. Elute proteins from agarose beads by adding 30 µL of 2X SDS-gel sample buffer and incubating for 30 min at room temperature (*see* **Note 8**).
17. Resolve proteins by SDS-PAGE and transfer to nitrocellulose membranes using a standard Western blot protocol (*13*).
18. The membrane is then blocked for 30 min in 10 mL of TBST containing 5% (w/v) nonfat dry milk.
19. The nitrocellulose is next incubated with 10 ml of TBST–5% milk containing 5 µg/mL of mouse monoclonal anti-FLAG M2 antibody for 1 h at room temperature.
20. Wash the nitrocellulose three times for 10 min each in TBST.
21. Incubate the nitrocellulose for 30 min with 10 mL of TBST–5% milk containing goat anti-mouse IgG conjugated to horseradish peroxidase at a dilution of 1:3000.

22. Wash the membrane five times for 10 min each in TBST.
23. Overlay the nitrocellulose with 1–2 mL of Supersignal chemiluminescence reagent for approx 5 min, allow the blot to drip-dry, wrap in plastic wrap, and visualize on x-ray film.
24. At this point blots can be treated with the denaturation solution to remove bound antibody and reprobed with the monoclonal HA antibody to detect receptor levels.

## 4. Notes

1. We typically examine receptor levels by Western blot analysis *(9,13)*, but other alternative methods can be used, such as radioligand binding analysis *(9)* or fluorescence-activated cell sorting (FACS) analysis *(8)*. One consideration that should be made is that when assessing total receptor levels by FACS analysis, the cells must be permeabilized so that all of the receptors in the cell are labeled. However, it is not known how effective labeling of intracellular receptor levels is or whether a receptor enters a compartment that is relatively impermeable. We have also examined agonist-promoted degradation of CXCR4 in CEM cells, an immortalized T-cell line *(9)*; therefore, this procedure is amenable to many different cell types.
2. We typically perform degradation assays using a single time point of 3 h. However, a time course of stimulation should be performed to establish the rate of CXCR4 degradation for a particular cell line.
3. We have found that 20-μL aliquots from 600 μL of lysates generated from subconfluent HeLa cells in 6-cm dishes results in an excellent CXCR4 signal as assessed by Western blot followed by chemiluminescence analysis. This must be determined empirically for each cell type and receptor being assessed.
4. We typically subject our samples to sonication before SDS-PAGE. Passing the lysate through a 27-gage needle 10 times can also be done in lieu or in addition to sonication.
5. We perform the ubiquitination assays in HEK293 cells that have been transiently transfected with an epitope-tagged version of CXCR4 (HA-tagged) and an epitope-tagged version of ubiquitin (3X FLAG) *(9)*. We have found that the degradation pathway for CXCR4 transiently transfected in HEK293 cells is similar to the pathway for degradation of endogenous CXCR4 in HeLa cells *(13)*. This procedure can easily be applied to examine CXCR4 ubiquitination by examining endogenous CXCR4 and ubiquitin.
6. In our system, cells are stimulated using a maximal concentration of SDF for 30 min. It may be prudent to perform dose–response and time-course assays to assess maximal ubiquitination in a given system.
7. We have found that under these conditions a maximal amount of epitope-tagged CXCR4 is immunoprecipiated from total cell lysates. However, the immunoprecipitation experiments must be optimized for any given system.
8. It is important that at this step the sample not be boiled. GPCRs readily aggregate when subject to high temperature, which will alter their migration on SDS-PAGE.

## References

1. Muller, A., Homey, B., Soto, H., et al. (2001) Involvement of chemokine receptors in breast cancer metastasis. *Nature* **410**, 50–56.
2. Thelen, M. (2001) Dancing to the tune of chemokines. *Nat. Immunol.* **2**, 129–134.
3. Belperio, J. A., Phillips, R. J., Burdick, M. D., Lutz, M., Keane, M., and Strieter, R. (2004) The SDF-1/CXCL 12/CXCR4 biological axis in non-small cell lung cancer metastases. *Chest* **125**, 156S.
4. Burger, M., Glodek, A., Hartmann, T., et al. (2003) Functional expression of CXCR4 (CD184) on small-cell lung cancer cells mediates migration, integrin activation, and adhesion to stromal cells. *Oncogene* **22**, 8093–8101.
5. Taichman, R. S., Cooper, C., Keller, E. T., Pienta, K. J., Taichman, N. S., and McCauley, L. K. (2002) Use of the stromal cell-derived factor-1/CXCR4 pathway in prostate cancer metastasis to bone. *Cancer Res.* **62**, 1832–1837.
6. Koshiba, T., Hosotani, R., Miyamoto, Y., et al. (2000) Expression of stromal cell-derived factor 1 and CXCR4 ligand receptor system in pancreatic cancer: a possible role for tumor progression. *Clin. Cancer Res.* **6**, 3530–3535.
7. Staller, P., Sulitkova, J., Lisztwan, J., Moch, H., Oakeley, E. J., and Krek, W. (2003) Chemokine receptor CXCR4 downregulated by von Hippel-Lindau tumour suppressor pVHL. *Nature* **425**, 307–311.
8. Li, Y. M., Pan, Y., Wei, Y., et al. (2004) Upregulation of CXCR4 is essential for HER2-mediated tumor metastasis. *Cancer Cell.* **6**, 459–469.
9. Marchese, A. and Benovic, J. L. (2001) Agonist-promoted ubiquitination of the G protein-coupled receptor CXCR4 mediates lysosomal sorting. *J. Biol. Chem.* **276**, 45,509–45,512.
10. Citri, A., Skaria, K. B., and Yarden, Y. (2003) The deaf and the dumb: the biology of ErbB-2 and ErbB-3. *Exp. Cell. Res.* **284**, 54–65.
11. Yu, D. and Hung, M. C. (2000) Overexpression of ErbB2 in cancer and ErbB2-targeting strategies. *Oncogene* **19**, 6115–6121.
12. Orsini, M. J., Parent, J. L., Mundell, S. J., Benovic, J. L., and Marchese, A. (1999) Trafficking of the HIV coreceptor CXCR4. Role of arrestins and identification of residues in the c-terminal tail that mediate receptor internalization. *J. Biol. Chem.* **274**, 31,076–31,086.
13. Marchese, A., Raiborg, C., Santini, F., Keen, J. H., Stenmark, H., and Benovic, J. L. (2003) The E3 ubiquitin ligase AIP4 mediates ubiquitination and sorting of the G protein-coupled receptor CXCR4. *Dev. Cell.* **5**, 709–722.

# 11

## Epithelial Cell Migration in Response to Epidermal Growth Factor

### Reema Zeineldin and Laurie G. Hudson

### Summary

Epidermal growth factor (EGF) is a ligand for the EGF receptor, a member of the erbB family of receptor tyrosine kinases. Activation of EGF receptor by EGF or other high-affinity ligands often results in increased migration of cells in physiological and pathological situations. There are numerous approaches for evaluating cell migratory response following EGF stimulation. Both qualitative and quantitative techniques will be presented in this chapter.

**Key Words:** EGF; EGF receptor; migration; wound-healing assay; time lapse; phagokinesis; electric cell-substrate impedance sensing (ECIS).

## 1. Introduction

Regulated cell movement is essential during embryogenesis, wound repair, angiogenesis, reproductive cycles, organ regeneration, and angiogenesis, whereas dysregulated cell migration is present during cancer progression, atherosclerosis, chronic wounds, and many inflammatory diseases *(1–3)*. Cell migration is a dynamic process that involves a complex cascade of events including activation of signal transduction cascades, regulation of cell–matrix and cell–cell adhesive contacts, and cytoskeletal reorganization. During locomotion, the cell becomes polarized and extends filopodia or lamellipodia in the direction of migration. This is accompanied by changes in cell morphology, cytoskeletal rearrangement, stabilization of attachments to the extracellular matrix enabling directed movement, and release of the trailing edge from the surface (reviewed in **refs.** *1,4*, and *5*). Each element of the process, from cell polarization to protrusion, adhesion formation, and, ultimately, rear retraction, involves multiple steps, complex signaling pathways, and spatial and temporal integration of numerous cellular functions.

From: *Methods in Molecular Biology, vol. 327: Epidermal Growth Factor: Methods and Protocols*
Edited by: T. B. Patel and P. J. Bertics © Humana Press Inc., Totowa, NJ

Cell migration can be stimulated by a variety of chemokines, growth factors, or extracellular matrix molecules. Epidermal growth factor (EGF) is a polypeptide growth factor that binds to EGF receptor, a member of the erbB receptor family of type 1 tyrosine kinases *(6)*. Activation of the EGF receptor is important for many diverse cellular functions, including growth and survival *(6)* and cell migration during development *(7,8)*, wound repair *(2,3)*, and cancer *(9)*. Thus, modulation of EGF-receptor-regulated cell migration may serve as an important target for therapeutic interventions in various disease states *(2,10)*.

A more detailed molecular understanding of the process of cell migration has been gained in recent years, and ongoing progress in the field is rapid. Because of the complexity of regulated cell movement, many distinct steps in the process may be studied, including regulation of signal transduction cascades, gene expression, protein–protein interactions, assembly of regulatory complexes, and other elements of migration. Discussion of each of these potential steps for analysis of cell migration is beyond the scope of this chapter. Rather, we will present qualitative and quantitative methods for evaluating functional migratory response to EGF with an emphasis on techniques that are readily accomplished in the typical laboratory. Qualitative methods include visualization of cells under phase-contrast microscopy in cell scattering or colony dispersion assays (**Figs. 1** and **2**), two-dimensional wound closure or in vitro wound-healing assays (**Fig. 3**), and time-lapse video microcopy. Each of these approaches provides direct, visual evidence for cell migration in response to EGF stimulation and following image acquisition, the results may be quantified using image analysis software.

Quantitative methods include phagokinesis assay (**Fig. 4**), various modifications of the Boyden chamber assay, and electric cell-substrate impedance sensing (ECIS). The phagokinesis assay *(11)* is based on production of a visible cell migration path due to cell phagocytosis of gold particles on the substratum. The approach allows the visualization of single cell movement and provides information on directionality. The Boyden chamber assay detects cells migrating through a porous membrane with or without extracellular matrix components. ECIS *(12,13)* is a real-time technique that is quantitative and completely automated and requires specialized equipment. With this method, cells are grown on a gold electrode and cell movement is detected by fluctuations in the electrode impedance (resistance and capacitance). This technique differs from the others presented in this chapter in that it is not limited by the wavelength of light and thus can detect cell motions of 1 nm (micromotion).

This chapter provides an introduction to approaches for the study of cell migration, but is not comprehensive. Many additional techniques can be found in Chapter 12 as well as elsewhere in the literature *(14–16)*.

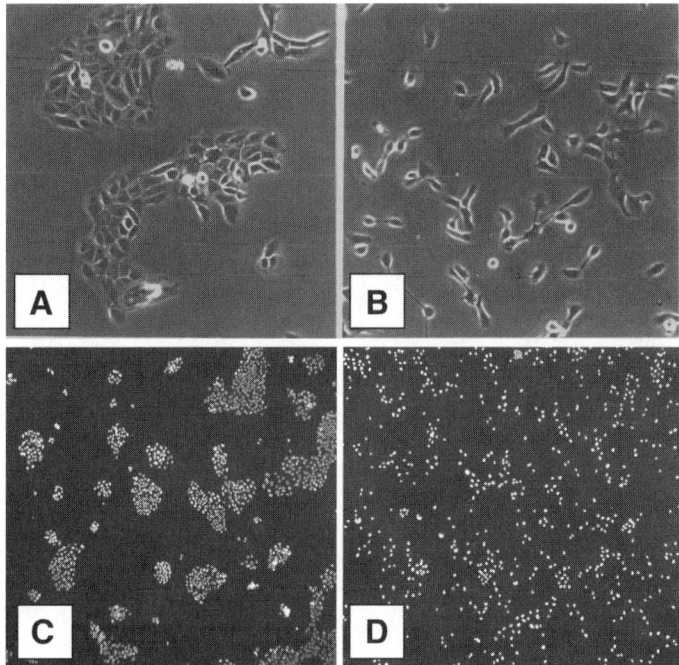

Fig. 1. Colony dispersion in response to epidermal growth factor (EGF). Cultures of a transformed human keratinocyte cell line (SCC12F) were maintained in serum-free medium containing 0.1% BSA (w/v) for 24 h prior to treatment without (**A,C**) or with (**B,D**) 10 n*M* EGF for 24 h. Phase-contrast microscopy (**A,B**) reveals dispersion (scattering) of cells in response to EGF and evidence of a fibroblastic morphology. DAPI staining and fluorescence microscopy (**C,D**) provides an alternate method for visualizing colony dispersion.

## 2. Materials

### 2.1. Cell Culture

1. Bovine serum albumin (BSA) can be purchased from Sigma (St. Louis, MO) and other major vendors.
2. Phosphate-buffered saline (PBS) can be obtained from Sigma and other major vendors.
3. Cell culture reagents can be purchased from Sigma and other major vendors.

### 2.2. Treatment With EGF

EGF can be purchased from Biomedical Technologies Inc. (Stoughton, MA), Upstate Biotechnology (Lake Placid, NY), and other vendors. Stock solutions of EGF may be made in base medium containing 0.1% (w/vol) BSA, then

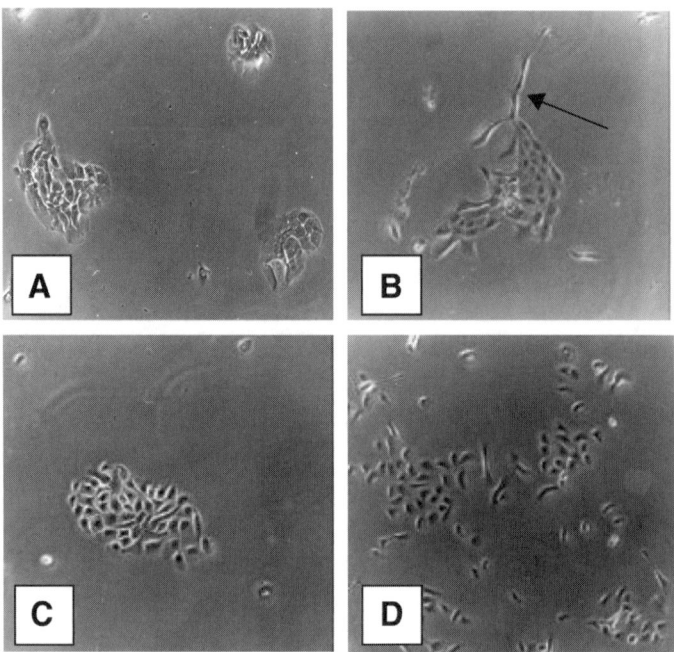

Fig. 2. Low-calcium colony dispersion assay. Cultures of normal human keratinocytes were maintained in defined keratinocyte growth medium (KGM) supplemented with 1 m*M* calcium (**A,B**) or basal (low-calcium) KGM (**C,D**) then treated without (**A,C**) or with (**B,D**) 10 n*M* EGF for 24 h. Under phase-contrast microscopy, changes in colony morphology are evident in EGF-treated cells (**B, arrow**), and upon disruption of cell:cell junctions under low-calcium conditions, EGF-dependent migration of cells is apparent (**D**).

aliquoted and stored frozen until needed. Once an aliquot is thawed, it may be stored at 4°C for up to 2 wk. Repeated freezing and thawing should be avoided.

## 2.3. Studies Involving Inhibition of the EGF Receptor

Tyrphostin AG 1478, which is a specific inhibitor of EGF receptor ($IC_{50}$ = 3 n*M*), may be purchased from Calbiochem (LaJolla, CA). Stock solutions may be made in dimethyl sulfoxide (DMSO) at a concentration of 20 μ*M*, then aliquoted and stored at –80°C until used. Once an aliquot is thawed, it can be stored at –20°C for re-use.

## 2.4. The Analysis of Phagokinesis

1. AuCl4H can be purchased from J.T. Baker Chemical Co. (Phillipsburg, NJ).
2. Sodium carbonate can be purchased from Sigma.
3. Paraformaldehyde can be obtained from EM Sciences (Gibbstown, NJ), and a solution of 0.1% (w/v) paraformaldehyde in water should be prepared fresh.

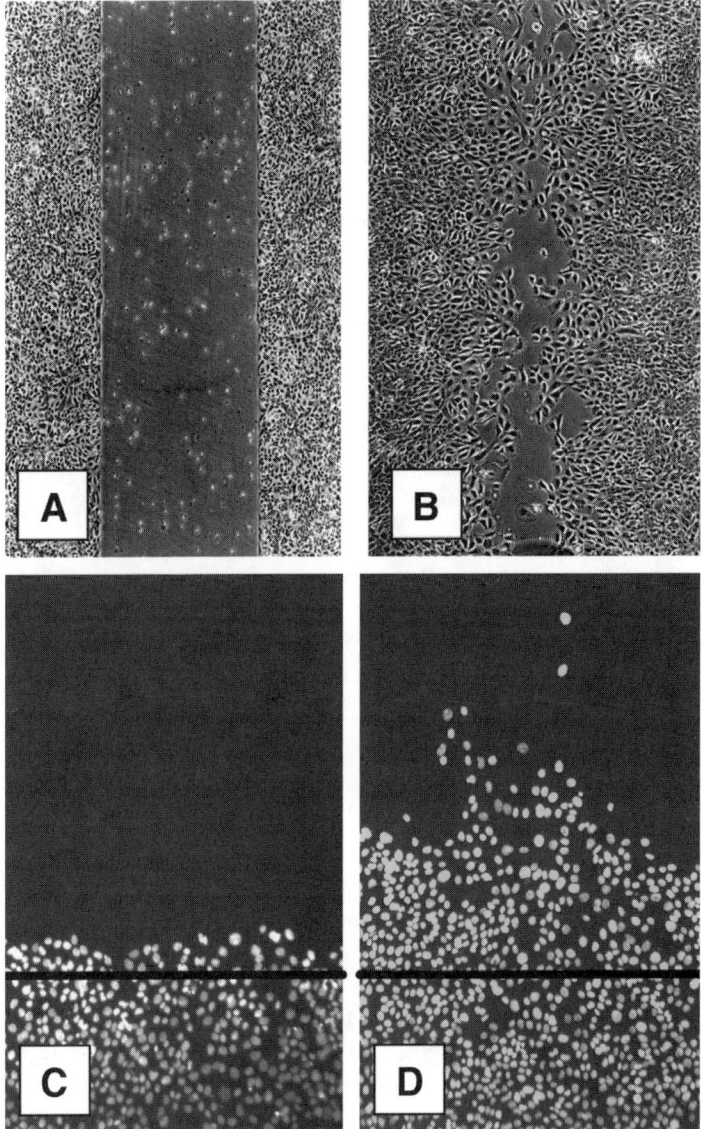

Fig. 3. In vitro wound closure in response to epidermal growth factor (EGF). Cultures of a transformed human keratinocyte cell line (SCC12F) were maintained in serum-free medium containing 0.1% BSA (w/v) for 24 h prior to treatment without (**A,C**) or with (**B,D**) 10 n*M* EGF for 24 h. Phase-contrast microscopy (**A,B**) reveals repopulation of the cleared area (in vitro wound closure) in response to EGF (**B**). DAPI staining and fluorescence microscopy (**C,D**) provides an alternate method for visualizing in vitro wound healing. The white line represents the margin of the original wound introduced in the cell monolayer. Methods for quantitating in vitro wound closure are described in the text.

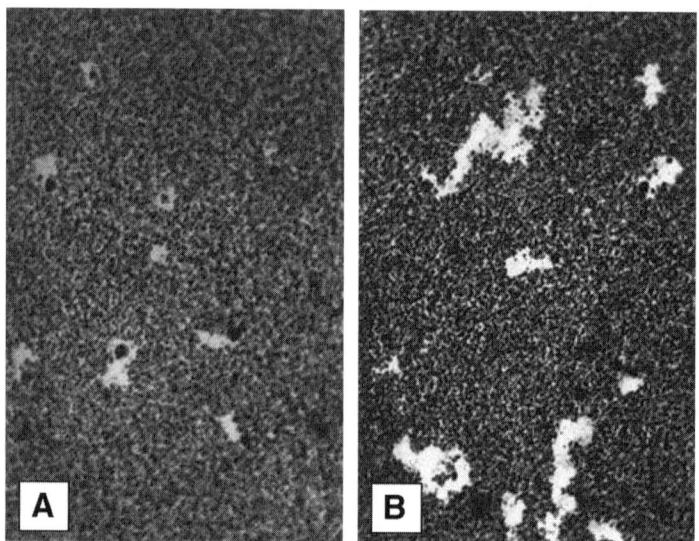

Fig. 4. Phagokinesis in response to epidermal growth factor (EGF). **(A)** Normal keratinocytes were plated on type I collagen substrate on colloidal gold-coated cover slips in keratinocyte growth medium. **(B)** Increase in phagokinetic track formation is evident in EGF-treated cultures.

## 2.5. Electric Cell-Substrate Impedance Sensing

All materials, equipment, and software for this assay can be purchased from Applied Biophysics (Troy, NY).

## 3. Methods
### 3.1. Cell Culture

1. Culture cells as recommended for the cell line to be tested and maintained in a humidified incubator at 37°C, 5% $CO_2$, unless otherwise directed.
2. For the colony-dispersion (cell-scattering) assay, the cells should be plated at low density and treated when colonies of more than 16 cells are established (*see* **Note 1**).
3. For in vitro wound closure assay, the cells should be grown to confluence prior to treatment.

### 3.2. Treatment With EGF

1. Remove growth medium.
2. Wash cells with PBS.
3. Serum-deprive cells by placing them into serum-free base medium containing 0.1% (w/v) BSA for 24 h (*see* **Note 2**).
4. Add EGF at a final concentration of 10 n$M$ (*see* **Note 3**).

### 3.3. Inhibition of EGF Receptor Activity

1. Prepare cells as described in **Subheading 3.2., steps 1–3**.
2. Treat cells with the EGF receptor catalytic inhibitor AG1478 at a final concentration of 2 μ$M$ or equal volume of DMSO for solvent controls.
3. After 30 min, treat solvent controls and AG1478-inhibited cultures without or with 10 n$M$ EGF (*see* **Note 4**).

### 3.4. Colony Dispersion Assay

1. Response to EGF may be determined qualitatively under phase-contrast microscopy (**Fig. 1A,B**). The morphological appearance of untreated cells should be compared with ligand-activated cells. Cells treated with EGF should display distinct morphological changes that may include (1) cell scattering in which cells migrate from the colony and a spindle-like appearance (fibroblastic morphology) with few cell–cell contacts (**Fig. 1B**), or (2) disruption of colony morphology without apparent migration of single cells from the colony (**Fig. 2B**).
2. Colony dispersion represents both cell migration and disruption of cell–cell contacts. In some cell lines, EGF treatment will not disrupt cell–cell junctions sufficiently to allow detachment of cells from neighboring cells, as shown in **Fig. 1B**. In this case, many epithelial cells may be placed in reduced calcium medium (<100 μ$M$ extracellular calcium; *see* **Note 5**) to disrupt cell–cell junctions *(17)* and treated as described above. Once junctions are disrupted, a migratory response may be evident (**Fig. 2C,D**).
3. Colony dispersion may be quantified by measuring average internuclear distance. Cells may be plated on chamber slides and treated as described above. Stain nuclei with 4',6'-diamidino-2-phenylindole (DAPI) (Vector Laboratories, Inc., Burlingame, CA) and mount cells. Visualize cells under fluorescence and obtain digital images with a charge-coupled device (CCD) camera (**Fig. 1C,D**). Quantitation of the average distance between the geometrical center of nuclei of adjacent cells can be measured using Openlab 3 software (Improvision, Lexington, MA) or other commercially available image analysis software. More than 100 measurements for each condition should be made to calculate each average of the internuclear distance. The statistical significance can be calculated using Student's *t*-test.

### 3.5. In Vitro Wound Closure Assay

1. Cells are grown to a confluent monolayer in six-well tissue culture plates.
2. A cell-free area is introduced into the cell monolayer by scraping the monolayer with a 1000 μL blue pipet tip (*see* **Note 6**). The location of the wound center should be marked on the underside of the plate as a reference point for image acquisition.
3. The cell monolayer is washed extensively with PBS to remove cell debris, and fresh serum-free medium containing BSA is added to each well. Cells are treated with or without EGF as described above.

4. Photographs of defined fields of the wound area relative to the reference point are obtained at this time ($t = 0$). Both edges of the wound should be visible in the photographic field to allow calculation and comparison of total area repopulated by cells over time.

5. Examine the cells under a phase-contrast microscope at various times to compare closure of the in vitro wound in control vs EGF-treated cells (**Fig. 3**).

6. To evaluate the relative contribution of cell migration to in vitro wound closure in the absence of cell proliferation, treat cells with a concentration of mitomycin C determined to inhibit mitogenesis in the cell line of interest before **step 2**.

7. Quantification of in vitro wound closure may be obtained in several ways. First, if the wound edge is marked on the bottom of the tissue culture plate, either (1) the number of cells migrating past the original wound edge per linear unit of the wound edge may be counted, (2) the average distance of cells from the original wound edge may be measured, or (3) the average distance between the geometrical center of nuclei of adjacent cells both within the confluent cell sheet and at the wound edge may be measured (**Fig. 3C,D**). Alternatively, total wound area may be calculated using image analysis software and compared to the initial wound area of defined fields (**Fig. 3A,B**).

### 3.6. Time-Lapse Video Microscopy of Individual Cells

1. Low density cell cultures are treated without or with EGF as described under **Subheading 3.2.**

2. Cells are monitored using a Panasonic AG-6730 analog S-VHS time-lapse recorder. Recorded images from a high-resolution Ikegami CCD video camera connected to an Olympus IX-50 inverted microscope, equipped with phase-contrast optics and an Olympus Peltier thermo-stated pad to maintain 37°C. Analyze cells at 37°C with a X40 semi-plan apochromatic phase-contrast objective.

3. Acquire movies at a time-frame rate of three frames every 10 s (real-time for a PAL system is 25 frames/s). This corresponds to an acquisition of 1125 frame/h, which leads to a 45-s time-lapse movie. When played and analyzed at a real-time frame rate (25 frames/s) these parameters of acquisition allow visualization of cell movement at about 80× real-time speed.

4. Express motility in µm/h and compare treated cells to untreated cells.

### 3.7. Assessing Phagokinesis

1. Preparation of gold particles and coating of cover slips (*11*):

   a. To 11 mL of distilled water add 1.8 mL of 14.5 m$M$ AuCl$_4$H solution and 6 mL of 36.5 m$M$ Na$_2$CO$_3$ solution.

   b. Heat in a glass beaker and, immediately after reaching the boiling point, add 1.8 mL of 0.1% (w/v) freshly prepared paraformaldehyde solution in water (*see* **Note 7**). The colloidal gold particles form within 1 min, producing a brownish solution that appears clear blue in transmitted light (*see* **Note 8**).

   c. Dip 22 × 22 mm square cover slips into a 1% (w/v) BSA solution prepared in distilled water and filtered through 0.20-µm filter. Drain the cover slip by touching its edge to a paper towel, then dip the cover slip into absolute ethanol and rapidly dry it using hot air-stream of a hairdryer.

e. Place the cover slip in a well of six-well plate, and layer it with 0.4 mL of the hot (80–90°C) gold particle suspension, and incubate for 45 min. Note that up to this point, no sterility precautions are required, but the following steps should be done under sterile conditions. Aspirate fluid, and wash the gold-coated cover slip several times with PBS. Transfer the cover slip into a new six-well plate and continue washing.

f. If desired, colloidal gold immobilized to cover slips may be coated with extracellular matrix components of choice.

2. Plate cells at a low density of approx 1000–2000 cells in six-well plates containing the gold-coated cover slips in 2 mL of growth media (*see* **Note 9**).

3. Allow the cells to attach (time will vary according to the cell line investigated), then treat one set of cover slips (thee cover slips) with EGF and leave the second set without treatment.

4. Incubate the cells for 18–20 h so that migration occurs, but for less time than the average cell doubling time.

5. Fix cells in 1% (w/v) formaldehyde/PBS.

6. The migrating cells displace gold salts, so migration is detected by presence of tracks (**Fig. 4**). Cell migration can be detected by visualization of areas free of the gold particles. A comparison of cell migration rates may be determined by measuring distances/time using a Nikon Diaphot inverted microscope and time-lapse photography or visualize by low power (20–40×) light microscopy under darkfield illumination. Quantification can be performed by computerized image analysis, which involves calculating the migration index (phagokinetic track area/cell) in multiple, nonoverlapping fields with a minimum of 50 cells analyzed (*see* **Note 10**). The results can be analyzed using Student's *t*-test.

## 3.8. Modified Boyden Chamber Assays

1. Cells are plated onto filter inserts (8 $\mu M$ pore size, BD Biosciences) and incubated at 37°C in humidified air with 5% $CO_2$ to allow for cellular adherence and spreading on the filter. The time will vary according to cell type and is likely to be 1 h or more.

2. Growth medium is replaced with serum-free medium containing BSA with or without EGF (*see* **Note 11**), and cells are placed into the humidified cell incubator (*see* **Subheading 3.1.**, **step 1**).

3. Cells are allowed to migrate through the filter. Optimum incubation times vary considerably for different cell types and will need to be determined empirically to distinguish between ligand-stimulated migration vs basal migration of unstimulated cells

4. After incubation, carefully remove the fluid from above the filter and from the bottom well.

5. Gently wipe the nonmigrated cells off with a cotton swab. Wipe twice using two swabs or both ends of a double-tipped swab. Remove the filter with forceps.

6. Stain the migrated cells with Diff-Quick (Fisher Scientific) on the bottom of the filter and count under a light microscope. Data can be expressed as a migration index (the number of cells that migrated in response to EGF relative to the number of cells that migrated in the absence of EGF) or percentage of the starting cell population that migrated during the incubation period if the migration period does not exceed cell doubling time (*see* **Note 12**).

## *3.9. Electric Cell-Substrate Impedance Sensing*

1. Cells are grown at $4 \times 10_4$ cells/well on small active gold electrodes (250 μm diameter) at the bottom of tissue culture wells (area 0.5 cm$^2$) in 400 μL of medium, which serves as an electrolyte.
2. Cells are treated in replicates as described in **Subheading 3.2.**, and the micromotion is recorded.
3. Apply a 1-μA, 4-kHz AC signal from a constant current source between the small electrode and a much larger counterelectrode (0.15 cm$^2$).
4. The voltage of the system is monitored by a lock-in amplifier, which can detect both magnitude and phase of the voltage appearing across the sample. The in-phase and out-of-phase voltages across the electrode are recorded by the lock-in amplifier once every second for measuring micromotion.
5. The ECIS software (Applied BioPhysics, Troy, NY) calculates the impedance (resistance and capacitance) values of the electrode over a designated period of time. As the cells move on the gold electrode, the sensitive nature of the lock-in amplifier detects the fluctuations in the resistance and capacitance values of impedance. These fluctuations are then statistically analyzed by the ECIS software, thus revealing the percentage variation in resistance, which in turn is a reflection of cellular micromotion on the electrode. The percentage variation for untreated cells in resistance is recorded. Treatment of cells with EGF would increase the fluctuations, indicating an increased motility, so the percentage variation in resistance is now elevated in comparison to controls.

## 4. Notes

1. Best results for quantitation by measuring internuclear distances (*see* **Subheading 3.4.**, **step 3**) are obtained when the assay is conducted with well-separated, small cell colonies.
2. If the cell line of interest is not viable in serum-free medium, use medium containing reduced serum (e.g., 1 or 0.5% serum by volume) instead of BSA.
3. The optimal concentration of EGF may need to be determined for the cell type of interest. Test concentration range between 0.1 and 100 n*M* EGF.
4. The effective concentrations of AG1478 for the cell line of interest should be determined by testing the ability of varying concentrations of AG1478 to block EGF-stimulated tyrosine phosphorylation using western blot analysis. Alternatively use an EGF receptor-blocking antibody (vendors include Upstate Biotechnology or Lab Vision Corporation, Fremont, CA).

5. Low-calcium medium may be generated as follows: add 30 g of Chelex resin (Biorad) to 500 mL of complete cell growth medium and mix at room temperature for 1 h. Separate Chelex resin from medium by low-speed centrifugation. Filter-sterilize medium and add sterile calcium chloride to 50 $\mu M$ final concentration and magnesium chloride to 1 m$M$ final concentration.

6. Use tips from the same vendor, apply equal pressure, and maintain a constant angle with the pipet tip to produce uniform size wounds. A new tip should be used for each wound.

7. If the paraformaldehyde solution is not prepared fresh, then the control over the colloidal form of gold and the thickness of the gold film would be lost.

8. If the solution appears purple-colored, this indicates that the particle distribution is not homogeneous, so do not use it.

9. If there is a high incidence of crossed or fused tracks, cells can be plated at a lower density.

10. To confirm viability of cells, one can do a trypan blue exclusion test.

11. Placement of EGF in the upper chamber will measure chemokinesis or increases in random migratory response. Placement of EGF in the lower chamber will measure chemotaxis, or movement toward a specific factor. A variation of the assay is the in vitro cell invasion assay using the same overall principles described above. Invasion assays rely on the ability of cells to migrate through extracellular matrix coated on the filter membrane. Precoated membranes are available through BD Biosciences.

12. Commercial migration assay kits based on fluorescence detection are now available as an alternate approach.

## Acknowledgments

This work was supported by National Institutes of Health grants R01DE12458 and R01AR42989.

## References

1. Ridley, A. J., Schwartz, M. A., Burridge, K., et al. (2003) Cell Migration: Integrating signals from front to back. *Science* 302, 1704–1709.

2. Singer, A. J. and Clark R. A. F. (1999) Cutaneous wound healing. *N. Engl. J. Med.* **341,** 738–746.

3. Hudson, L. G. and McCawley, L. J. (1998) Contributions of the epidermal growth factor receptor to keratinocyte motility. *Micros. Res. Tech.* **43,** 444–455.

4. Lauffenburger, D. A. and Horwitz, A. F. (1996) Cell migration: a physically integrated molecular process. *Cell* **84,** 359–369.

5. Pollard, T. D. and Borisy, G. G. (2003) Cellular motility driven by assembly and disassembly of actin filaments. *Cell* **112,** 453–465.

6. Jorissen, R. N., Walker, F., Pouliot, N. Garrett, T. P. J., Ward, C. W., and Burgessa, A. W. (2003) Epidermal growth factor receptor: mechanisms of activation and signaling. *Exp. Cell Res.* **284,** 31–53.

7. Moghal, N. and Sternberg, P. W. (2003) The epidermal growth factor system in Caenorhabditis elegans. *Exp. Cell Res.* **284,** 150–159.

8. Shilo, B.-Z. (2003) Signaling by the Drosophila epidermal growth factor receptor pathway during development. *Exp. Cell Res.* **284,** 140–149.

9. Holbro, T., Civenni, G., and Hynes, N. E. (2003) The ErbB receptors and their role in cancer progression. *Exp. Cell Res.* **284,** 99–110.

10. Arteaga, C. L. (2003) ErbB-targeted therapeutic approaches in human cancer. *Exp. Cell Res.* **284,** 122–133.

11. Albrecht-Buehler, G. (1977) The phagokinetic tracks of 3T3 cells. *Cell* **11,** 395–404.

12. Giaever, I. and Keese, C. R. (1991) Micromotion of mammalian-cells measured electrically. *Proc. Natl. Acad. Sci. USA* **88,** 7896–7900.

13. http://www.biophysics.com

14. Maheshwari, G., Wiley, H. S., and Lauffenburger, D.A. (2001) Autocrine epidermal growth factor signaling stimulates directionally persistent mammary epithelial cell migration. *J. Cell Biol.* **155,** 1123–1128.

15. Ware, M. F., Wells, A., and Lauffenburger, D. A. (1998) Epidermal growth factor alters fibroblast migration speed and directional persistence reciprocally and in a matrix-dependent manner. *J. Cell Sci.* **111,** 2423–2432.

16. Mataraza, J. M., Briggs, M. W., Li, Z., Entwistle, A, Ridley, A. J., and Sacks, D. B. (2003) IQGAP1 promotes cell motility and invasion. *J. Biol. Chem.* **278,** 41,237–41,245.

17. Volberg, T., Geiger, B., Kartenbeck, J., and Franke, W. W. (1986) Changes in membrane-microfilament interaction in intercellular adherens junctions upon removal of extracellular calcium ions. *J. Cell Biol.* **102,** 1832–1842.

# 12

# Motility Signaled From the EGF Receptor and Related Systems

**Alan Wells, Brian Harms, Akihiro Iwabu, Lily Koo, Kirsty Smith, Linda Griffith, and Douglas A. Lauffenburger**

## Summary

Cell motility is now recognized as central to many biological processes. Growth factors, such as those that activate the epidermal growth factor receptor (EGFR), drive biochemically and biologically distinct subsets of migration critical for (neo)organogenesis and tumor invasion. Thus, modulation of these events requires an understanding of the controls of EGFR-mediated motility. Deconstruction of motility into its component events enables this deeper insight. Herein we describe methods that measure the overall motility and its parameters as well as the biophysical processes extension, de-adhesion/retraction, and contraction.

**Key Words:** Lamellipodia; de-adhesion; contraction/contractility; extension; migration.

## 1. Introduction

The epidermal growth factor receptor (EGFR) and its ligands have been implicated in a number of physiological (branching morphogenesis and wound repair) and pathological (carcinogenesis and tumor progression) states *(1–4)*. These phenomena have been demonstrated in animal models of both transgenic mammals and drosophila. However, the bases of these organismal developments are the underlying cellular responses elicited by EGFR activation. The two main cellular responses studied are proliferation and migration, though in select cell types EGFR signaling has more diverse effects, including cellular differentiation, trans/dedifferentiation, and metabolic regulation.

Ligand binding to EGFR elicits a pleiotropic intracellular signaling response. Numerous cell-signaling cascades are activated; most of these are common to other receptors with intrinsic tyrosine kinase activity. Some of the signaling

From: *Methods in Molecular Biology, vol. 327: Epidermal Growth Factor: Methods and Protocols*
Edited by: T. B. Patel and P. J. Bertics © Humana Press Inc., Totowa, NJ

cascades appear operative in only some cell types; for example, PI3 kinase activity is readily noted in carcinoma cells but is triggered only weakly if at all in fibroblasts. Other well-studied cascades, such as those via ERK mitogen-activated protein (MAP) kinase and phospholipase C (PLC)γ, are strongly stimulated in all situations examined. These numerous intracellular biochemical events that occur downstream of EGFR activation have been discussed in many recent reviews *(5–8)*. The activation status of most of these signaling pathways can be probed by standard detection assays that are addressed under those particular molecules.

Many, but not all, of these signaling pathways have been linked to cellular responses. Recently, interest in studying cell motility has increased. Herein we will detail methods for studying EGFR-mediated cell motility and its component aspects: lamellipod protrusion, cell contractility, and substratum de-adhesion.

Cell motility is simply productive locomotion of the entire cell from one point to the next. However, a large number of parameters is involved, primarily cell speed, persistence, and dispersion, each of which carries it own implications and can distort mass "movement" assessments. Thus, cell motility is best evaluated by tracking individual cells *(9)*, but because of the time commitment and complexity of obtaining such data, we will also describe a rapid, high-throughput way for testing relatively motility. There are conceptual concerns as to whether cell motility should be measured in two or three dimensions. Many cells, mainly epithelial types, do migrate in vivo in a quasi-two-dimensional space, whereas others, mainly mesenchymal cells, move through three-dimensional spaces (although they do appear to move with a ventral and dorsal surface rather than transversely symmetrically *[10]*). Further, movement through matrices, or pores in Boyden chambers, requires cell deformation and/or matrix remodeling, which, while important for translocations in vivo, are properties distinct from motility *per se*. Thus, such assays do not strictly determine motility unless the cells are actually tracked during the process. For these reasons we will present protocols for measuring cell motility in two dimensions.

Cell motility can be broken into three main steps that can be conceived of being iterated, though they do occur simultaneously. First, the cell must extend a lamellipod; second, the cell exerts a contractile force to pull the cell body and tail forward; and last, the trailing edge must deattach. We present quantitative methods for each of these subprocesses.

## 1.1. Cell Motility

Determination of cell motility requires timed measurements, during which the cells move. Two common methods are utilized. One, the in vitro wound healing model, relies on population dynamics, while the other, single-cell tracking, images individual cells. The former has the advantage of "scoring" many cells simultaneously and high throughput; these attributes promote the use of the in vitro wound-healing assay as a "screen" or first pass at cell motility. The latter process, single cell tracking, provides more information on the attributes of motility (speed, persistance, etc.), but this high granularity limits throughput. Recent advances in automated image tracking software promise to combine the positive attributes on both assays *(11)*, but these developments are nascent and will not be described herein.

The two assays described track cell motility across a surface. Such an approach is supported by the fact that many adherent cells (mainly epithelial cells) move over such surfaces (though these may be folded) and that even when cells move through three-dimensional spaces, they often behave as if there are defined ventral and dorsal surfaces *(10,12)*. Other assays enumerate the ability of cells to move across or through barriers. These assays, which include the Boyden chamber, transmigration through extracellular matrices, and single-cell tracking during invasion in the matrix, measure not only motility but also cell deformability and, depending on the assay ability to modify the matrix. As such these assay are not pure motility assays, and will not be described herein.

## 1.2. Lamellipod Extension

Forward movement is driven by lamellipod protrusion. The molecular motors that underlie the force generation are being deciphered in quantitative detail; controlling mechanisms include the Rho GTPases directing enucleation of actin filaments by the Arp2/3 complex at the cell front *(14–16)*. This is regulated by EGF receptor activation of PLCγ to hydrolyzed PIP2 at the cell front, liberating actin-binding proteins to sever, cap, and nucleate actin *(17,18)*, and PI3-kinase to generate novel protein docking sites, possibly to reinforce lamelllipod directionality *(16)*. Still, these exciting biochemical cascades exist to drive forward protrusion, whose rate and direction cannot be derived from the constituent events but needs to be described as an integrated whole. Direct visualization of lamellipod protrusion is the only way to truly evaluate this phenomenon at present. This can be captured simultaneously with motility during single cell tracking or, if more detail is needed, at higher magnification for protrusion analysis in isolation.

New techniques purport to isolate and collect lamellipodia, either by enticing extension into small-bore pipet tips or, for greater collection volume, through low μm-sized pores *(19,20)*. While these newer techniques hold promise for understanding molecular localizations during lamellipod protrusion, they may not isolate the broad fronts that are noted during keratocyte-like or macrophage locomotion. As these are still in their infancy, they will not be described herein.

## 1.3. Contractile Force Generation

During motility, intracellular contractile forces are needed to both bring the cell body forward and retract/detach the tail. The necessary contractility is thus dictated by the adhesiveness of the substratum. However, the force generation is not simply across the cell, since this contractility must still enable lamellipod protrusion. For this reason it is likely that contractility is asymmetric and compartmentally localized *(22)*. For instance, the contractile force appears not to extend into the actual lamellipod but rather from the base of the lamellipod *(23,24)*, although some have noted traction from the front of dominant lamellipodia *(25)*. At the rear of the cell, contractile forces must extend to the end, as such pulling results in membrane shedding (in fibroblasts and epithelial cells) when moving across adhesive surfaces *(26–28)*. Curiously, while cells move with a ventral/dorsal orientation, at least in fish keratocytes, the contractile forces appear similar in respect to these surfaces *(29)*.

Compartmentalization of molecular motors may underlie the asymmetry in contractile force generation. Most of the contractile force for motility is derived from acto-myosin motors *(30,31)*. One model has myosin activation, as determined by phosphorylation of myosin light chain (MLC), being in a gradient with the highest concentration in the rear *(32–34)*. A second posits differential localization of myosin isoforms leading to contractile asymmetry *(29,35–37)*. These models are not exclusive, and it is likely that integrative studies with molecular probes and specific interventions will define both being operative during locomotion.

Cell contractile forces are measured indirectly by determining their perturbations on matrix external to the cells. One caveat for all these assays is that they rely on transmission of the forces through transmembrane linkages to the external reporters. As these linkages are dynamic, being modified by both mechanical and biochemical processes, and often in asymmetrical fashion, most measurements will underestimate the actual forces generated within the cell *(1,38)*. Still, these assays have yielded insights into force generation both across and within a cell. The simplest assay is the gel compaction assay, in which a collagen-based matrix is embedded with cells, released from external constraints, and then challenged; the change in gel diameter or weight (as water is extruded) determines the extent of contractile forces exerted on the matrix. A more challenging technique measures tension generated in a isometric manner; the cell-populated, collagen-based matrix is fixed at two ends. A third approachable method visualizes movement of substratum as a cell moves across the surface. This technique generates pliable supports (silicone or acrylamide) and visualized physical deformation or displacement of markers embedded in the supports. This latter method provides information at the cellular and subcellular level, whereas the two former methods determine population responses.

Newer developments aim towards creating μm-sized force detectors that will provide individual point tension measurements at the subcellular level *(39,40)*; however, these are highly specialized and technically challenging being limited to few labs and will not be discussed herein.

### 1.4. De-Adhesion

Detachment of the trailing edge of the cell defines translocational motility. In the absence of such release, cells can be stimulated to extend and pull their nuclei forward, but ultimately recoil to the original site *(44)*. Adherent cells accomplish this process differently, depending on the adhesiveness to the subsubstrate *(45)*. Integrins disengage from substratum during movement across low adhesive surfaces, while movement across highly across surfaces results in membrane shedding *(26,27)*. The intracellular limited proteinase calpain cleaves components of the integrin/adhesion complex (although the critical elements are to be determined) to weaken the adhesiveness to the substrate *(1)*. This appears to occur asymmetrically with episodic activation of calpain at the rear to enable retraction (during fibroblastoid-like motility) and calpain-assisted adhesion remodeling and turnover at the front of the cell to enable progressive extensions. It is quite possible that these two modes of adhesion regulation utilize different calpain isoforms and target different adhesion components.

Even though adhesiveness to the substrate is being decreased at many points in the cell, the cell does not actually detach from the surface. Rather, the adhesiveness is quantitatively decreased, necessitating quantitative measurements of the cell–substratum interaction. Further, despite the spatio-temporal nature of this lessened adhesion, current technology only allows for whole cell measurements. These can be approached two ways. Cells can be challenged by forces perpendicular to the plane of attachment to assess adhesiveness. Alternatively, shear flow can be used to detach cells; this mode may be most appropriate for endothelial cells that experience shear forces in vivo, but may also introduce variables such as cell profile and shear signaling.

## 2. Materials

### 2.1. Cell Motility Measurements

1. Six-well tissue culture plates, with or without substrate coated.
2. Rubber policeman.
3. 1X phosphate-buffered saline (PBS), sterile.
4. Quiescence medium, namely, restricted serum conditioned medium that does not support growth but maintains viability of the cells being tested.
5. Other reagents, e.g., growth factors, cytokines, chemokines, inhibitors, etc.
6. Inverted microscope.

7. Digital acquisition and processing system, namely, charge-coupled device (CCD) camera with controller, computer, display monitor, and image analyzing software.

## 2.2. Single-Cell Tracking

1. Cell culture dishes for experiments.
2. Protein or extracellular matrix (ECM) solution for surface modification.
3. 1% bovine serum albumin (BSA)/PBS solution, sterile.
4. 1X PBS, sterile.
5. Ethylenediaminetetraacetic acid (EDTA) solution for releasing cells from culture plates.
6. Quiescence medium, namely, restricted serum conditioned medium that does not support growth but maintains viability of the cells being tested.
7. Motility medium: hydroxyethyl piperazine ethane sulfonate (HEPES)-buffered for microscope systems without $CO_2$ control.
8. Other reagents, e.g., growth factors, cytokines, chemokines, inhibitors, etc., for motility medium.
9. Mineral oil.
10. Inverted microscope with automated image acquisition system, namely, low magnification phase-contrast or differential interference contrast (DIC) objective (5–10×), computer controlled X,Y,Z-axis microscope stage control, and CCD camera.
11. Heated microscope stage insert for maintaining constant temperature of cell dish during tracking.
12. Suitable image acquisition and image analysis software.

## 2.3. Measuring Lamellipod Extension

Similar equipment is used as with single-cell videotracking. A higher-magnification microscope objective may be used (10–20×).

## 2.4. Gel Compaction Assay

1. Cell culture media containing fetal bovine serum (FBS).
2. Quiescence medium (serum-free media that retains cell viability but lacks growth factors/ECM proteins that are being tested) containing 1 mg/mL BSA.
3. 24-well tissue culture plates.
4. Scalpel for matrix release.
5. Trypsin/EDTA solution.
6. Collagenase solution.
7. PBS.
8. BSA.
9. FBS.
10. Type 1 rat tail collagen.

11. NaOH neutralizing media (0.1 *N* in quiescence media).
12. Clinical/cell centrifuge.
13. Test agonists/antagonists:

> Cell ligands (growth factors, ECM proteins).
> Pharmacological inhibitors.

## 2.5. Measurements of Isometric Force Generated

1. Type I collagen.
2. Force transducer (model 52-9545, Harvard Apparatus, South Natick, MA).
3. Stepper motor.
4. 15-mm-diameter cylindrical Teflon molds.
5. Organ bath.

## 2.6. Measurement of Cell Detachment (De-Adhesion)

1. 24-well tissue culture plates.
2. Protein or ECM solution for surface modification.
3. BSA.
4. 1X PBS, sterile.
5. HEPES buffer, pH 7.4.
6. Quiescence medium, namely, restricted serum conditioned medium that does not support growth but maintains viability of the cells being tested.
7. Other reagents, e.g., growth factors, cytokines, chemokines, inhibitors, etc.
8. Enzyme-linked immunosorbent assay (ELISA) sealing tape.
9. Plate centrifuge, e.g., Beckman CS6R.
10. Phase-contrast inverted microscope.
11. Digital acquisition and processing system, namely, CCD camera with controller, computer, display monitor, and image-analyzing software.

## 2.7. Shear Force Detachment Assay

1. Parallel-plate shear flow detachment chamber—typical components include:

   a. A base plate with an observation slot.
   b. A gasket or spacer that defines the height of the chamber.
   c. A cover plate with an observation slot.
   d. Inlet and outlet tubing for cell, reagents, and shearing medium.

2. ECM proteins (optional).
3. BSA.
4. 1X PBS.
5. HEPES buffer, pH 7.4.
6. Shearing medium/buffer (e.g., $Mg^{2+}$ and $Mn^{2+}$ containing PBS).
7. Infusion pump.
8. Inverted microscope.

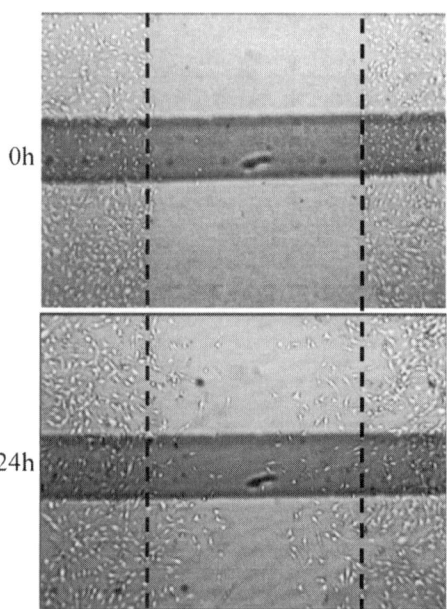

Fig. 1. Example of a wound-healing assay performed with NR6 WT cells stimulated with EGF (1 n*M*). The dashed lines denote the original edges of the wound at time 0. Note the marker line on the plate bottom that enables reorientation for picture acquisition and consistency of gap measurement.

## 3. Methods

### 3.1. Measurement of In Vitro Wound Healing (Fig. 1)

This assay is utilized for high throughput evaluation of cell motility under numerous conditions. This can be across different surfaces (or surface density), multiple stimuli, or many inhibitors. Advantages are ease of experimentation and accessibility of means of evaluation. The assay does not measure motility *per se*, but forward dispersion; as such, quasi-unidirectionality is enforced by movement from a wound front (in which reverse or even lateral movement is limited by cell–cell contact). Thus, if the cells in question do not readily form near confluent monolayers, this assay is not appropriate.

1. Add 2 mL of cell suspension to each well of six-well plates at $0.5–3.0 \times 10^5$ cells/mL depending on time required to reach confluence (or plate cells on six-well plates and grow to confluence in growth medium).
2. After 1–3 d, quiesce confluent monolayer cells with quiescence medium for appropriate time (or switch to quiescence medium prior to reaching desired confluence).
3. Aspirate medium and wash the cells once with PBS.

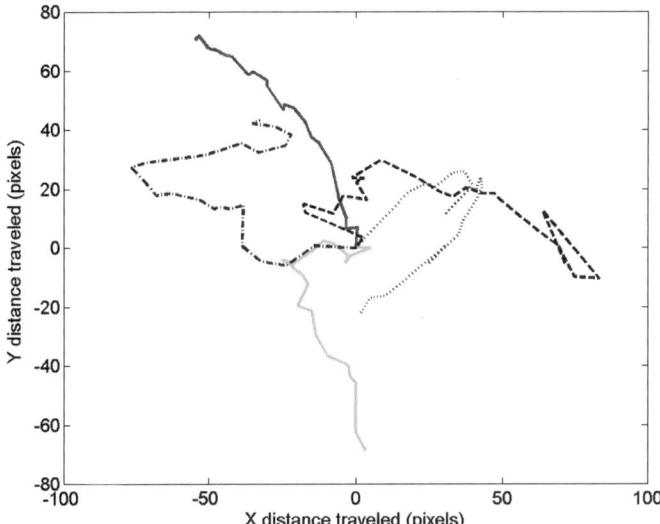

Fig. 2. Single-cell tracks traced by time-lapse photography. Five cells were tracked by following the centroid over a 24-h period (15-min intervals) by time-lapse photography. Each cell was then plotted with time 0 at the origin of the grid. From these tracks one can determine total path length for speed and dispersion as well as persistence time and persistance path length.

4. Scrape the cells with a rubber policeman to introduce a linear acellular area at the center of the well.
5. Wash with PBS three times to remove the scraped-off cells and then add 2 mL of quiescence medium to each well.
6. Take digital photograph of plates at 0 h at a magnification that captures both sides of the "wound" (and mark on the plate bottoms where photographs are taken); this is usually 20× or 40× magnification.
7. Incubate with or without EGF (1–25 n*M*) in the presence or absence of inhibitors, cytokines, or chemokines for 24 h at 37°C (*see* **Note 1**).
8. Acquire digital photographs at times up to and exceeding 24 h.
9. To determine the distance cell migrated into the acellular area, measure the distance between the two boundaries of cells on the photographs of same magnification.

This assay can be used for soluble factors in co-culture system *(13)* with the bottom well being used as above and the upper insert holding the producing cells of question.

### 3.2. Monitoring of Single-Cell Tracking (Fig. 2)

Single-cell tracking enables accurate measurement of cell speed and its individual attributes such as persistence and dispersion. Furthermore, at appropriate magnification, the individual stages of protrusion, retraction, and lamellipodia switching can be evaluated (discussed later). However, current technology and analysis software is limiting in being expensive and best used when customized to individual utilization. Additionally, this assay evaluates only a few dozen cells during any single assay run, limiting throughput.

1. Surface preparation: coat migration assay culture dish with protein or extracellular matrix according to individual protocol. A sample protocol for fibronectin (Fn) is as follows (*see* **Note 2**):

   a. Sterilize culture dish using ethanol (plastic dish) or ultraviolet (UV) (glass) followed by rinsing with sterile PBS.
   b. Add 1 mL of sterile Fn/PBS solution to dish and coat overnight (4°C) or for 1 h (37°C). Rinse with 2× 1 mL PBS.
   c. Block surface with 1 mL 1% BSA/PBS solution for 1 h (37°C). Rinse with 2× 1 mL PBS.
   d. The dish is now ready for use. If using immediately, keep dish surface under PBS and warm in incubator; alternatively, dish may be stored for 24 h under PBS at 4°C.

2. Cell preparation: place cells in $G_0$ state to minimize cell proliferation during cell tracking (*see* **Note 3**).

   a. Passage cells such that they are in log phase growth at the onset of serum starvation. Confluency of approx 50% before quiescence of cells is desirable.
   b. Quiesce cells for the appropriate time using serum-restricted medium.
   c. Use EDTA solution for release of cells from culture dish while preserving integrin membrane proteins for cell adhesion.
   d. (Optional) Suspend cells for 1 h in quiescence medium to reduce adhesion-mediated cell signaling to baseline levels.
   e. Centrifuge cells and resuspend in motility medium. Plate cells onto migration dish at approx 50 cells/mm² and allow to adhere and develop steady-state behavior. A layer of mineral oil should be overlaid over the motility medium to prevent evaporation from the migration dish; for some migration systems, a specialized dish cover may be used. Bicarbonate-based media will additionally require a perfusion system or $CO_2$-based microincubator system integral to the microscope; HEPES-buffered medium does not require such a system but is sometimes not tolerated well by cells.

3. Image acquisition and movie analysis:

   a. Run the appropriate computer program for acquisition of the X,Y,Z positions of selected image fields on stage. Ten to twenty separate fields on the migration dish are desirable with multiple single cells visible at each position (*see* **Note 4**).
   b. Run automated image capture macro for times up to and exceeding 24 h. For instance, for NR6 fibroblasts, a picture of each image field is captured every 15 min for 12 h. An electronic shutter is desirable for limiting the light exposure of the cells when not capturing images. The experimentalist's choice of

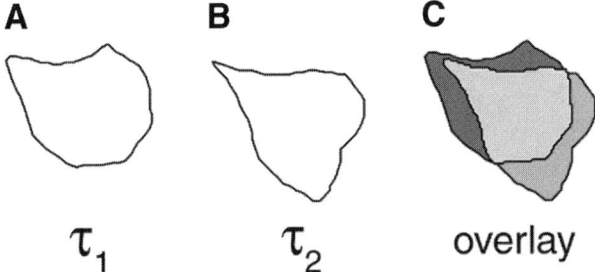

Fig. 3. Outline of extended lamellipodia. The cells were outlined at two time points from time-lapse photography. By overlaying the two images, one can calculate both the area extended at the front and the area retracted at the rear.

total observation time and time interval between successive images will be dictated by ease of data analysis, average cell speeds, and the cells' long-term steady-state behavior.

c. Sort images into digital movies. Commercial and custom software packages are available for outlining cell perimeters for determination of cell centroid positions over time. Multiple software packages are also available for analysis of these cell paths and generation of motility parameter data.

### 3.3. Monitoring of Lamellipod Tracking (Fig. 3)

Lamellipod protrusion determines both the rate of forward locomotion and the direction of the cell motility. Direct visualization enables the determination of both the frequency of extension and the cell fraction protruded within a defined time period; these two parameters determine cell speed *(21)*. Simultaneously, lamellipod directionality and generation of alternative lamellipodia can be assessed, with these events directly dictating persistence and dispersion. The main drawback to this method is its labor intensity and low throughput.

1. Lamellipodial tracking experiments may be performed similarly to single-cell videotracking experiments, with slight modifications. Cells may be observed under greater magnification, and total observation times can be reduced to approximately 1–4 h.
2. Cell outline traces from subsequent cell images are overlaid. The areas of protruding regions of the cell are measured; they signify absolute lamellipodial protrusion rate. Normalizing this area to the area of the entire cell gives a measure of specific membrane extension rate (*see* **Note 5**).

### 3.4. Gel Compaction Assay

This high-throughput assay can be seen to integrate tension transmitted to the matrix overtime. The hydrated gel is free is be reshaped by such transmitted forces. Contractile forces will extrude water, and this will not be regained upon release of tension, such as occurs during motility. The transient tensions transmitted to the matrix during rapid cell movement will incrementally compact the matrix over time. Thus, this is not a direct measure of contractile force (as transmitted to the external environment), but rather a time-integrated measure of multiple contractions.

### 3.4.1. Inhibitor Pretreatment

1. Harvest cells from monolayer culture using 0.25% trypsin/EDTA.
2. Neutralize cell/trypsin solution with culture media, spin at approx $500g$ for 6 min, resuspend pellet in quiescent media at a concentration of $1 \times 10^6$ cells/mL.
3. Incubate ± inhibitors for 30–60 min prior to gel preparation.

### 3.4.2. Hydrated Collagen Matrix Preparation

1. Centrifuge the cells at 1000 rpm for 6 min, resuspend the pellet in quiescent media at a concentration $4 \times 10^6$ cells/mL.
2. Prepare neutralized collagen/cell solution as follows, adding ligands and inhibitors where appropriate. Add solutions in the order listed (A–D) (also *see* **Note 6**).

| Total volume | Volume/ matrix | A. Culture media ± test ligands and inhibitors | B. NaOH (0.1 $N$) | C. Collagen (4 mg/mL) | D. Cells ($4 \times 10^6$/mL) |
|---|---|---|---|---|---|
| 3 mL | 0.5 mL | 1.25 mL | 0.25 mL | 0.75 mL | 0.75 mL |

The final collagen concentration is 1 mg/mL and the cell concentration is $1 \times 10^6$ cells/mL.

3. Add 0.5 mL of collagen/cell solution to each of 5 wells of a 24-well plate (*see* **Note 7**).
4. Leave matrices to polymerize at 37°C for 60 min.
5. Overlay each matrix with 1 ml of quiescent media ± test inhibitors
6. Gently release matrices from sides and bottom of well using a scalpel.
7. Leave matrices to contract for 24 h.
8. Weigh matrices.

### 3.4.3. Matrix Digestion to Determine Cell Viability After 24 H

1. Rinse gels in PBS for 10 min.
2. Add three gels/test group to a 15-mL tube, treat with 750 μL of 0.05% trypsin at 37°C for 10 min.
3. Add 750 μL of collagenase solution (0.5 mg/mL, pH 7.2) to each tube at 37°C for 10 min, mix well by pipeting to assist digestion.
4. Add 400 μL of FBS per tube to quench collagenase activity.
5. Centrifuge samples for 6 min at 1000 rpm.
6. Add 400 μL of culture media and 400 μL of trypan blue to each pellet, mix well, and assess cell viability using a hemocytometer.

## 3.5. Measurement of Isometric Force Generation

By securing the gel, initial contractile force generation can be measured. In such a setting, only forces aligned in the direction of the tensometers will be scored; this alignment is dependent on both the vector of the force generation as well as the cell's orientation. The tensometers detect the release of transmitted tension, such as occurs during rear retraction (41). Thus, this assay, though of low throughput, provides information complementary to the above gel compaction assay.

1. Suspend cells (approx $10^6$) and type I collagen (0.75 mg, in solution) in 1 mL of tissue-culture medium for each experiment.
2. Pour suspension into the annular space of cylindrical Teflon mold (1 mL/mold).
3. Allow to gel by incubation at 37°C in 5% $CO_2$ for 1 h.
4. Remove the ring-shaped gels from the molds and placed in tissue culture medium for approx 40 h. In order to maintain the original diameter, place on sterile parallel spacing rods to prevent the matrix contraction due to cell spreading from collapsing the ring-shaped gel diameter during incubation.
5. Mount gel vertically on force transducer (see **Note 8**). Loop the tissue ring over both a rod connected to a computer-controlled stepper motor and a hook suspended from the force transducer.
6. Place mounted gel in an organ bath containing 50 mL of HEPES-buffered, serum-free DMEM thermostatted at 37°C. Adjust rod position to its original length (15 mm).
7. Permit gel to reach an equilibrium force over a period of 30–60 min.
8. Expose gel to desired treatment (see **Note 9**). Contraction force is monitored by computer data acquisition from the force transducer.

### 3.5.1. Protocol to Measure Surface Deformation

The preceding two assays measure population responses. Visualization of the effects of cells upon underlying substratum would assess individual cell attributes. This has been enabled by methods using pliable substrata upon/into which adhesion molecules have been incorporated (42,43). Imaging the deformation of these pliable substratum, or displacement of markers embedded within them, allows for calculated force generation. High-resolution imaging provides data on the both vector and degree of forces at a subcellular level. As this requires specialized equipment and analyses, the precise protocols will not be provided herein. However, if necessary, one should contact those versed in the techniques for instruction and collaborations.

## 3.6. Monitoring Cell Detachment (De-Adhesion)

### 3.6.1. Inverted Centrifugation

Forces perpendicular to the plane of attachment hold the possibility of determining adhesive forces in isolation from cellular geometry and adaptive signaling. Cells are subjected to centrifugation in an inverted manner, generating a force that dissociates them from the substratum while minimizing shear. An advantage of this method, in addition to minimizing active signaling and adaptive mechanisms, is its technical ease. Additionally, when examining multiple events, conditions, or regulators, the throughput can be quite high; though different detachment forces require cumbersome separate runs.

1. Coat 24-well plates with protein or extracellular matrix according to individual protocol (This step is optional.) For Amgel (*see* **Note 10**) or Matrigel this is as follows:

   a. Add 0.5 mL of Amgel or Matrigel solution (0.5 µg/mL) and 0.5 m*M* HEPES to each well and allow to sit for 1 h at room temperature.
   b. Aspirate the solution and wash once with 1X PBS.
   c. For blocking, add 0.5 mL of 1% BSA solution (in PBS) to each well and allow to sit for 1 h at room temperature.
   d. Aspirate the BSA solution and wash twice with 1X PBS.

   For fibronectin (Fn) or vitronectin (Vn) coating, protocol is as follows:

   e. Incubate each well with Fn or Vn solution (0.1 mg/µL to 100 µg/mL in PBS) for 2 h at 37°C or overnight at 4°C.
   f. Remove the protein solution and rinse twice with PBS.
   g. Block substrates with BSA solution (1–4% in PBS) for 30 min to 1 h at 37°C.
   h. Remove the BSA solution and rinse twice with PBS.
   i. Store wells under PBS at 4°C until use.

2. Plate cells at a concentration of $10^5$ cells/mL with quiescent medium into 24-well plates and incubate for 12–24 h at 37°C.
3. Treat with or without EGF (1–25 n*M*) in the presence or absence of inhibitors, cytokines, or chemokines for appropriate time.
4. Count the number of cells on the plates by phase-contrast inverted microscopy (best to acquire digital image).
5. Fill the plates completely with basal medium with 1% BSA and 25 m*M* HEPES and seal them with ELISA sealing tape (*see* **Note 11**).
6. Centrifuge the plates in inverted position for 5–10 min at 1500–3000*g* at 37°C using a plate centrifuge. If a control plate is used, keep it inverted (1 g) for the duration of the centrifugation (*see* **Notes 12–14**).
7. After the centrifugation, count the number of cells remaining on the plates by phase-contrast inverted microscopy (best to acquire digital image) (*see* **Notes 15** and **16**).
8. For each experimental condition, an adhesion-strength profile can be constructed as the percentage of cells remaining adherent versus the centrifugal force. If a linear profile is obtained, the force required to detach approximately 50% of the initially seeded cells can be used as a measure of cell adhesion strength for that particular experimental condition.

## 3.6.2. Shear Force Detachment

A second method for quantifying adhesiveness involves cell detachment from the substratum by shear forces of aqueous solutions. Disadvantages of this method are that the shear itself may actively regulate adhesion to the substratum, the geometry of the cells partially determines the detachment flow, and, if the cells are cohesive, the shear tends to "peel" the monolayer. However, the assay, if performed in a Hele-Shaw chamber, can provide analogue information as the detachment forces needed in a single run. Also *see* **Note 17** for other methods of shear force detachment.

1. Coat a glass cover slip with an ECM protein of interest (optional) (proceed as under **Subheading 3.6.1.**).
2. Seed cells over cover slip with quiescent medium and incubate for 12–24 h at 37°C.
3. Treat with or without EGF (1–25 n*M*) in the presence or absence of inhibitors, cytokines, or chemokines for desired length of time.
4. Begin to construct the flow chamber (*see* **Note 18**).
5. Add shearing medium (*see* **Note 19**) to the flow area defined by the base plate and the gasket until a slight meniscus forms.
6. Sandwich the cell-seeded slide between the gasket and the cover plate with the cells facing the flow area.
7. Secure the apparatus for a tight seal.
8. Place the chamber onto an inverted microscope and connect tubing to the reservoir of shearing medium.
9. Introduce a slow flow to remove dead cells and debris.
10. Image cells from selected fields for an initial cell count.
11. Initiate flow.
12. At the termination of shear flow, image cells from the previously chosen fields for a final cell count (*see* **Note 20**).

## 4. Notes

1. Mitomycin-C (0.5 µg/mL for most fibroblasts) can be used to determine mitogenesis-independent cell migration.
2. Cell adhesion and spreading during the initial period after cell plating, while another aspect of cell motility, are largely transient and properly separated from subsequent cell translocation. Also, the experimentalist must empirically determine the time-course of activation (relative to steady-state behavior) for cytokines, growth factors, inhibitors, and so on, under study.
3. Proliferation during tracking experiments is undesirable since observations of multiple cell types reveal that motility behavior immediately after cell division is altered relative to that of nonproliferating cells. Thus cell division reduces the already limited pool of cells available for data analysis.

4. Single-cell migration behavior is highly dependent upon cell confluency and cell–cell contact. In addition, highly confluent cells are difficult to gently separate for use in subsequent single-cell experiments.

5. Older implementations of this method relied on manual overlay of cell outlines; current software packages such as DIAS (Solltech, Inc, Oakdale, IA) allow automated overlay and calculation of protrusion rate, thus vastly increasing data throughput.

6. When preparing the collagen/cell solution it is important to make extra solution to allow for transfer loss; for example, 3 mL of solution is prepared for five 0.5-mL gels. Mix the solution well by pipetting, before casting the gels, to assure an even distribution of cells.

7. Prepare the gels as quickly as possible to prevent polymerization beginning before gels are cast. Do not make more than 3 mL of matrix solution at a time and the collagen must be at 4°C when added.

8. Minimizing contracture of the gel prior to mounting is critical to obtaining reproducible measurements.

9. Treatment with agents and inhibitors often requires concentrations significantly higher than those published for monolayer effects. This is due to diffusion into the gel. If the inhibitor is essentially inactive on its own, it is suggested to incorporate monolayer concentrations into the solution from which the fast-protein liquid chromatography (FPLC) is made at time of gelation.

10. Amgel is an ECM suspension derived from human amniotic membrane that contains a variety of ECM proteins including collagen I and IV, fibronectin, tenascin, and laminin. Unlike Matrigel, there is no growth factor present. Amgel in solution provides a relatively easy way to coat desired surfaces with a complex biologically active substrate.

11. When filling the wells with medium prior to sealing with the ELISA tape, overfill slightly to form a meniscus in order to prevent air bubbles from forming underneath the ELISA tape. Air bubbles that come in contact with the cells when plates are inverted can remove or kill the cells and inflate the extent of cell detachment.

12. Centrifugation speed is chosen empirically as the force required to detach approximately half of EGF-treated cells.

13. If necessary, one can stack two plates per carrier in the centrifuge; more than two is not recommended due to differences in g-force (can be calculated on radius but hard to recreate for independent replication) and lack of full stability of the plates.

14. Due to safety and operation related considerations, the force ranges allowed in most commercially available supercentrifuges are limited and may be insufficient to detach well-adherent cells in the absence of EGF treatment.

15. Each condition should be in four- to eightfold replicate as a result of a loss of cells if there are any bubbles in individual wells.

16. It is best to acquire images both pre- and post-shear force for later enumeration, as extended time periods in both cases introduce experimental variability (i.e., released cells may settle by gravity and reattach, etc.).

17. Other methods of shear force detachment include radial-flow chamber and spinning disk, which generates differential force fields radially.

18. For parallel-plate chamber assay, throughput in shear force can be improved through geometric manipulation (e.g., tapered channel width or height). In general, the wall shear stress is proportional to the flow rate and viscosity and inversely proportional to the channel width and squared height.

19. Increased shear forces can be achieved by increasing shearing medium viscosity using an inert agent (glycerol, etc.).

20. It is important to ensure that only laminar flow is applied for readily characterizable shear force; cells at the chamber edges and within the entrance length, where the steady state flow is not fully developed, are not fully reliable markers of detachment.

## References

1. Glading, A., Lauffenburger, D. A., and Wells, A. (2002) Cutting to the chase: calpain proteases in cell migration. *Trends Cell Biol.* **12,** 46.
2. Normannol, N., Bianco, C., DeLuca, A., and Salomon, D. S. (2001) The role of EGF-related peptides in tumor growth. *Frontiers Biosci.* **6,** 685.
3. Wells, A. (2000) Tumor invasion: role of growth factor-induced cell motility. *Adv. Cancer Res.* **78,** 31.
4. Curtiss, J., Halder, G., and Mlodzik, M. (2002) Selector and signalling molecules cooperate in organ patterning. *Nat. Cell Biol.* **4,** E48.
5. Carpenter, G. (2000) The EGF receptor: a nexus for trafficking and signaling. *Bioessays* **22,** 697.
6. Schlessinger, J. (2002) Ligand-induced, receptor-mediated dimerization and activation of EGF receptor. *Cell* **110,** 669.
7. Wells, A. (2002) The EGF receptor signaling system. a model for growth factor receptor signaling, in *Hormone Signalling* (Goffin, V. and Kelly, P. A., eds.), Kluwer Academic Publishers, Norwell, MA, p. 57.
8. Schlessinger, J. (2000) Cell signaling by receptor tyrosine kinases. *Cell* **103,** 211.
9. Lauffenburger, D. A. and Horwitz, A. F. (1996) Cell migration: a physically integrated molecular process. *Cell* **84,** 359.
10. Friedl, P., Zanker, K. S., and Brocker, E.-B. (1998) Cell migration strategies in 3-D extracellular matrix: differences in morphology, cell matrix interactions and integrin function. *Microsc. Res. Tech.* **43,** 369.
11. Demou, Z. N. and McIntire, L. V. (2002) Fully automated three-dimensional tracking of cancer cells in collagen gels: determination of motility phenotypes at the cellular level. *Cancer Res.* **62,** 5301.
12. Friedl, P. and Wolf, K. (2003) Tumour-cell invasion and migration: diversity and escape mechanisms. *Nat. Rev. Cancer* **3,** 362.
13. Satish, L., Yager, D., and Wells, A. (2003) ELR-negative CXC chemokine IP-9 as a mediator of epidermal-dermal communication during wound repair. *J. Invest. Dermatol.* **120,** 1110.
14. Ridley, A. J. (2001) Rho family proteins: coordinating cell responses. Trends Cell Biol. 11, 471.
15. Pollard, T. D. (2003) The cytoskeleton, cellular motility, and the reductionist agenda. Nature 422, 741.

16. Condeelis, J. (2001) How is actin polymerization nucleated in vivo? *Trends Cell Biol.* **11**, 288.
17. Chou, J., Beer-Stolz, D., Burke, N., Watkins, S. C., and Wells, A. (2002) Distribution of gelsolin and phosphoinositol 4,5-bisphosphate in lamellipodia during EGF-induced motility. *Int. J. Biochem. Cell Biol.* **34**, 776.
18. Chou, J., Burke, N. A., Iwabu, A., Watkins, S. C., and Wells, A. (2003) Directional motility induced by EGF requires cdc42. *Exp. Cell Res.* **287**, 47.
19. Brahmbhatt, A. A. and Klemke, R. L. (2003) ERK and RhoA differentially regulate pseudopodia growth and retraction during chemotaxis. *J. Biol. Chem.* **278**, 13,016.
20. Wyckoff, J. B., Segall, J. E., and Condeelis, J. S. (2000) The collection of the motile population of cells from a living tumor. *Cancer Res.* **60**, 5401.
21. Ware, M. F., Wells, A., and Lauffenburger, D. A. (1998) Epidermal growth factor alters fibroblast migration speed and directional persistence reciprocally and in matrix-dependent manner. *J. Cell Sci.* **111**, 2423.
22. Schmidt, C. E., Horwitz, A. F., Lauffenburger, D. A., and Sheetz, M. P. (1993) Integrin-cytoskeletal interactions in migrating fibroblasts are dynamic, asymmetric and regulated. *J. Cell Biol.* **123**, 977.
23. Lee, J., Leonard, M., Oliver, T., Ishihara, A., and Jacobson, K. (1994) Traction forces generated by locomoting keratocytes. *J. Cell Biol.* **127**, 1957.
24. Svitkina, T. M., Verkovsky, A. B., McQuade, K. M., and Borisy, G. G. (1997) Analysis of the actin-myosin II system in fish epidermal keratocytes: mechanism of cell body translocation. *J. Cell Biol.* **139**, 397.
25. Burton, K., Park, J. H., and Taylor, D. L. (1999) Keratocytes generate traction forces in two phases. *Mol. Biol. Cell* **10**, 3745.
26. Regen, C. M. and Horwitz, A. F. (1992) Dynamics of b1 integrin-mediated adhesive contacts in motile fibroblasts. *J. Cell Biol.* **119**, 1347.
27. Friedl, P., Maaser, K., Klein, C. E., Niggemann, B., Krohne, G., and Zanker, K. S. (1997) Migration of highly aggressive MV3 melanoma cells in 3-D collagen lattices results in local matrix reorganization and shedding of a2 and b1 integrins and CD44. *Cancer Res.* **57**, 2061.
28. Jay, P. Y., Pham, P. A., Wong, S. A., and Elson, E. L. (1995) A mechanical function of myosin II in cell motility. *J. Cell Sci.* **108**, 387.
29. Galbraith, C. G. and Sheetz, M. P. (1999) Keratocytes pull with similar forces on their dorsal and ventral surfaces. *J. Cell Biol.* **147**, 1313.
30. Wessels, D., Murray, J., Jung, G., Hammer, J. A., and Soll, D. (1991) Myosin Ib null mutants of Dictyostelium exhibit abnormalities in movement. *Cell Motil. Cytoskel.* **20**, 301.
31. Mitchison, T. J. and Cramer, L. P. (1996) Actin-based cell motility and cell locomotion. *Cell* **84**, 371.
32. Post, P. L., DeBiasio, R. L., and Taylor, D. L. (1995) A fluorescent protein biosensor of myosin II regulatory light chain phosphorylation reports a gradient of phosphorylated myosin II in migrating cells. *Mol. Biol. Cell* **6**, 1755.

33. Verkhovsky, A. B., Svitkina, T. M., and Borisy, G. G. (1999) Self-polarization and directional motility of cytoplasm. *Curr. Biol.* **9,** 11.
34. Matsumura, F., Ono, S., Yamakita, Y., Totsukawa, G., and Yamashiro, S. (1998) Specific localization of serine 19 phosphorylated myosin II during cell locomotion and mitosis of cultured cells. *J. Cell Biol.* **140,** 119.
35. Conrad, A. H., Jaffredo, T., and Conrad, G. W. (1995) Differential localization of cytoplasmic myosin II isoforms A and B in avian interphase and dividing embryonic and immortalized cardiomyocytes and other cell types in vitro. *Cell Motil. Cytoskel.* **31,** 93.
36. Chung, C. Y., Potikyan, G., and Firtel, R. A. (2001) Control of cell polarity and chemotaxis by Akt/PKB and PI3 kinase through the regulation of PAKa. *Mol. Cell* **7,** 937.
37. Saitoh, T., Takemura, S., Ueda, K., et al. (2001) Differential localization of nonmuscle myosin II isoforms and phosphorylated regulatory light chains in human MRC-5 fibroblasts. *FEBS Lett.* **509,** 365.
38. Galbraith, C. G. and Sheetz, M. P. (1998) Forces on adhesive contacts affect cell function. *Curr. Opin. Cell Biol.* **10,** 566.
39. Galbraith, C. G. and Sheetz, M. P. (1997) A micromachined device provides a new bend on fibroblast traction forces. *Proc. Natl. Acad. Sci. USA* **94,** 9114.
40. Tan, J. L., Tien, J., Pironte, D. M., Gray, D. S., Bhadriraju, K., and Chen, C. S. (2003) Cells lying on a bed of microneedles: an approach to isolate mechanical force. *Proc. Natl. Acad. Sci. USA* **100,** 1484.
41. Allen, F. D., Asnes, C. F., Chang, P., Elson, E. L., Lauffenburger, D. A., and Wells, A. (2002) EGF-induced matrix contraction is modulated by calpain. *Wound Repair Regen.* **10,** 67.
42. Beningo, K. A. and Wang, Y.-L. (2002) Flexible substrata for the detection of cellular traction forces. *Trends Cell Biol.* **12,** 79.
43. Burton, K. and Taylor, D. L. (1997) Traction forces of cytokinesis measured with optically modified elastic substrata. *Nature* **385,** 450.
44. Shiraha, H., Glading, A., Chou, J., Jia, Z., and Wells, A. (2002) Activation of m-calpain (calpain II) by epidermal growth factor is limited by PKA phosphorylation of m-calpain. *Mol. Cell. Biol.* **22,** 2716.
45. Palecek, S., Huttenlocher, A., Horwitz, A. F., and Lauffenburger, D. A. (1998) Physical and biochemical regulation of integrin release during rear detachment of migrating cells. *J. Cell Sci.* **111,** 929.

# 13

## Methods for Determining the Proliferation of Cells in Response to EGFR Ligands

Gregory J. Wiepz, Francis Edwin, Tarun Patel, and Paul J. Bertics

### Summary

The evaluation of cell proliferation can be accomplished by several methods. The number of cells can be determined directly by counting manually (e.g., hemocytometer) or automatically (e.g., Coulter counter or flow cytometer). The amount of DNA, which reflects the number of cells or the stage of the cell cycle, can be quantified by incorporation of labeled nucleotides (e.g., [³H]-thymidine) or nucleic acid stains (e.g., propidium iodide). Alternatively, the relative metabolic activity, which is correlative with the number of cells, can be determined through the use of metabolic dyes and measurement of the colored metabolites (e.g., MTT and MTS). Each assay has its advantages and limitations. Determining which assay to use will depend on the equipment available, the experimental design, and the questions being addressed. In this chapter we will describe methods for the use of a hemocytometer, [³H]-thymidine incorporation, cell cycle analysis with propidium iodide by flow cytometry, and evaluation of cellular metabolic activity with the MTS reagent.

**Key Words:** Hemocytometer; Coulter counter; flow cytometer; [³H]-thymidine; propidium iodide; MTT; MTS; cell cycle; cell proliferation.

## 1. Introduction

An important action of epidermal growth factor (EGF) is its ability to induce cell proliferation in numerous cell types *(1)*. Cell proliferation is essential for normal physiological development as well as wound repair *(2)*. However, when the mechanisms that regulate cellular proliferation become dysfunctional, the inability to regulate growth leads to diseases that include tumor development and cancer. Following the binding of EGF to the EGF receptor, a number of distinct intracellular signaling pathways are activated, which leads to expression of proteins necessary to initiate the cell cycle and begin the produc-

From: *Methods in Molecular Biology, vol. 327: Epidermal Growth Factor: Methods and Protocols*
Edited by: T. B. Patel and P. J. Bertics © Humana Press Inc., Totowa, NJ

tion of DNA. Understanding the mechanisms that regulate cell proliferation, as well as the intracellular machinery that is involved in cell cycle progression, would permit better management of certain disease states.

The cell cycle has been described as having four separate stages *(3)*. The initial stage is referred to as G0 or G1. This stage follows the last mitosis but preceeds the cell becoming committed to advancement through the cell cycle. If a cell is not proliferating, then it resides in G0 stage and is considered quiescent. As the cell enters and advances through the G1 phase, usually under the control of a growth factor such as EGF, it begins to prepare the necessary machinery that will be required to duplicate the DNA during the S phase. Following DNA replication and the completion of S phase, the cells enter the G2 phase, where the required proteins are prepared for the execution of the mitosis or the M phase. The M phase is where the duplicate chromosomes are separated and partitioned and the cell undergoes cytokinesis to produce two identical daughter cells.

The effect of EGF or other treatments on cell proliferation can be quantified by various assays that utilize cell counting, DNA quantification, or substrate conversion based on metabolic activity. Cell counting can be accomplished by the use of a hemocytometer, Coulter counter, or flow cytometer. The choice of technique depends on the equipment that is available and the number of samples. Quantification of the amount of DNA can be performed in different ways depending on the question being addressed and the equipment available. One common method quantifies the incorporation of [$^3$H]-thymidine into proliferating cells. The [$^3$H]-thymidine is modified to [$^3$H]-thymine and incorporated into newly synthesized DNA. Following cell lysis, the amount of incorporated radioactivity can be determined by scintillation counting. This measurement is an indication of the amount of DNA newly synthesized due to treatments. A second method for DNA quantification employs the use of DNA stains detectable by spectroscopy or flow cytometry. Propidium iodide (PI) is a common DNA stain used for assessing cell cycle progression. When the cells are fixed and stained with PI and evaluated on a flow cytometer, the analysis will characterize the amount of DNA per cell based on the amount of fluorescence per cell. Analysis software can then determine the number of cells that are in the specific stages of the cell cycle (G1, S, or G2/M).

The third method for detecting cell proliferation is the use of metabolic dyes, such as 3-[4,5-dimethylthiazolyl-2]-2 5-diphenyltetrazolium bromide (MTT). When introduced to live cells, the compound is converted to an insoluble metabolite, formazan, which can be quantified by absorbance readings at 490 nm because the relative enzymatic activity is consistent among cells. The amount of conversion is directly related to the number of live cells present. Recent developments have led to a modification of the MTT tetrazolium compound such that the metabolite is a soluble formazan (3-[4,5-dimethylthiazolyl-2-yl] 5-[3-carboxymethoxyphenyl]-2-[4-sulfophenyl]-2H-tetrazolium, inner salt; MTS) and does not require a lengthy solubilization step like that required in the MTT methods. One such assay is the CellTiter 96 Aqueous Non-Radioactive Cell Proliferation Assay (Promega Corp, Madison, WI).

## 2. Materials

### 2.1. Cell Counting in Hemocytometer Chamber

1. Tissue culture dishes and the appropriate tissue culture media.
2. Serum starvation media: normal tissue culture media without serum and supplemented with 0.1% bovine serum albumin (BSA).
3. 0.25% Trypsin/ethylenediaminetetraacetic acid (EDTA) solution (2.5% trypsin solution: Media Tech; cat. no. 25054).
4. Treatments.
5. Hemocytometer (Hausser Scientific, Fisher,VWR).
6. Trypan blue dye solution (Invitrogen, cat. no. 15250-061).
7. 70% Ethanol.
8. Microscope: upright scope with 10× objective lens.

### 2.2. Analysis of Proliferation by [³H]-Thymidine Incorporation

1. 24-well tissue culture plates and the appropriate culture media for the specific cell type being studied.
2. Serum starvation media: normal tissue culture media without serum and supplemented with 0.1% BSA.
3. Sterile 1X phosphate-buffered saline (PBS).
4. Treatments (e.g., epidermal growth factor [EGF], Heregulin, transforming growth factor [TGF]-α, and so on).
5. [³H]-thymidine (Perkin Elmer, Amersham-Pharmacia).
6. 10% Trichloroacetic acid (TCA; Sigma T-6399; prepared as a 100% solution in water).
7. Solubilization solution: 0.1% sodium dodecyl sulfate (SDS) (from a 10% stock; Sigma L-4390) and 0.1 $N$ NaOH (from a 1 $N$ stock; Sigma: S-8045).
8. Scintillation fluid.
9. Liquid scintillation counter.

### 2.3. Analysis of Proliferation Through Cell Cycle Stage Analysis

1. Tissue culture dishes (60–100 mm) and the appropriate tissue culture media.
2. Serum starvation media: normal tissue culture media without serum and supplemented with 0.1% BSA.
3. Treatments (e.g., EGF, heregulin, TGF-α, and so on).
4. Trypsin/EDTA solution.
5. Beckman J6-HC or similar centrifuge.
6. 1X ice-cold PBS.

7. RNAse (Sigma, cat. no. R-5503 or any other source).
8. Propidium iodide (Aldrich, cat. no. 287075 or any other source).
9. Flow cytometer.

### 2.4. Determining Cell Proliferation by Analyses of Metabolic Activity as Estimated by the Bioreduction of Tetrazolium Compounds to Formazan

1. Tissue culture multiwell plates: 12, 24, 48, or 96 wells.
2. The appropriate tissue culture media with serum.
3. Serum starvation media: normal tissue culture media without serum and supplemented with 0.1% BSA.
4. Treatments (e.g., EGF, Heregulin, TGF-$\alpha$, and so on).
5. MTS/PMS solution (CellTiter 96 Aqueous Non-Radioactive Cell Proliferation Assay; Promega Corporation, Madison, WI).
6. Absorbance plate reader: capable of reading at 490 nm wavelength.

## 3. Methods

### 3.1. Analysis of Cell Proliferation by Cell Counting in a Hemocytometer

Counting cells in a hemocytometer is the simplest and most direct method of quantifying the number of cells in solution. This method also allows determining the percentage of viable cells of a cell population by employing trypan blue dye exclusion. This procedure is adaptable for adherent cells as well as cells in suspension and requires a minimal amount of equipment. In brief, following treatment, the cells are harvested and an aliquot of cells is combined with trypan blue dye solution and loaded into the hemocytometer chamber. The cells are then viewed on microscope and the number of live cells present within a specific area are counted. Because the volume of the chamber is precisely defined (commonly 0.1 mm³), the number of cells counted per area multiplied by the dilution factor will determine the number of live cells per milliliter.

1. Plate cells for each condition at a confluency of 30–40% in the appropriate size dish in normal media and grow overnight at the appropriate incubation conditions (*see* **Note 1**).
2. Aspirate media and supplement with serum-free media containing 0.1% BSA, and incubate overnight at the appropriate incubation conditions.
3. Following the specific experimental design, add treatments to the cells and continue the incubation under appropriate conditions for the specific time points. Usually, for this assay, the measurements are taken over the next few days (*see* **Note 2**).
4. To count the cells, lift the cells by trypsinization if they are adherent (*see* **Notes 3** and **4**).
5. Gently resuspend the cells throughout medium by pipetting up and down.
6. Combine an aliquot of cell suspension (50 µL) with an equal volume of trypan blue solution (50 µL).

7. Clean hemocytometer and cover slip using 70% ethanol.
8. Place cover slip squarely on top of hemocytometer, lightly moistening the polish surface of the slide by breathing on it before pressing the cover slip in position.
9. Load hemocytometer with a small amount of cell solution (~15 µL) to both sides so that the fluid entirely covers the polished surface of each chamber but does not overflow (*see* **Note 5**).
10. Using a 10× objective of the microscope, locate the primary square millimeter. Count the number of cells in each square millimeter (*see* **Notes 6** and **7**).
11. Calculate the average number of cells per square millimeter by dividing the total number of cells by the number of large squares counted.
12. Calculations for the number of cells per mL = average number of cells counted per square millimeter × dilution factor × 10,000 (conversion of 0.1 mm³ to 1 mL).

## 3.2. Proliferation Analysis of Adherent Cells by [³H]-Thymidine Incorporation

Cells can convert [³H]-thymidine into [³H]-thymine, which can then be incorporated into DNA as it is synthesized. This procedure allows the quantification of only newly synthesized DNA and provides a direct reflection of the effect of any treatments that maybe affecting early cell cycle progression *(4)*. Briefly, quiescent cells in G0 of the cell cycle are exposed to treatments in the presence of [³H]-thymidine. After an appropriate incubation period (usually 12–18 h) the cells are washed, and the cellular products are precipitated with TCA and then solublized to allow efficient scintillation counting. Finally, the amount of [³H] present is determined by scintillation counting. This value can then be compared between treatments directly. It is important to understand that you are measuring only newly synthesized DNA. Once the cells have completed the S phase, the value will not change until the next S phase.

1. Plate cells in 24-well plates in the appropriate serum containing media (*see* **Note 8**).
2. Culture the cells overnight, and then replace the growth media with serum-free media containing 0.1% BSA as the protein source.
3. Culture the cells overnight to allow them to become quiescent and enter into G0 phase of the cell cycle.
4. The cells can then be exposed to the required treatments supplemented with 1 µCi of [³H]-thymidine (*see* **Note 8**).
5. Continue the incubation for 18–24 h.
6. At the end of the incubation, place the plates on ice.
7. Aspirate the [³H]-thymidine containing medium and wash three times with ice-cold PBS (*see* **Note 9**).
8. Incubate the cells at 4°C with ice-cold 10% TCA for 10 min. Use a sufficient volume to cover the cells.
9. Rinse the plates briefly three times with ice-cold PBS (*see* **Note 9**).
10. Add 100 µL of 0.1% SDS in 0.5 *N* NaOH to each well to solublize the precipitated proteins/DNA and incubate at room temperature for 2 h or overnight.

11. Solubilize the cells by pipetting up and down several times and aliquot 50 μL from each well into 5 mL of scintillation fluid and count for [³H] on a liquid scintillation counter.

12. Direct comparisons can be made between treatments based on the amount of [³H] incorporated. Additionally, 10 or 20 μL from the remaining lysate can be used to determine the protein concentration using the bicinchoninic acid (BCA) method (Micro BCA protein assay kit from Pierce or any other source). The counts can then be normalized using the protein concentration of each sample.

### 3.3. Analysis of DNA Content and Cell Cycle Progression by Flow Cytometry

Analysis of the DNA content of a cell can provide a great deal of information about the cell cycle and consequently the effect of added stimuli (e.g., transfected genes, drug treatment, growth factors) on the cell cycle progression *(4)*. The flow cytometer has the distinct advantage of being able to evaluate each cell individually and record the amount of fluorescent label contained within each cell. Thus, the amount of dye incorporated into the DNA will be reflective of the amount of DNA present. By analyzing the amount of propidium iodide per cell, the whole population can be defined as to the number of cells that are in each stage of the cell cycle at that specific time. Briefly, the cells are plated, serum starved, and treated. After an appropriate incubation time (12–24 h), the cells are fixed with ethanol, treated with RNase to remove any double-stranded RNA, and then labeled with propidium iodide, which binds to double-stranded DNA. The cells are then analyzed on a flow cytometer, and the data analysis is performed by a program such as ModFit LT (Verity Software House, Topsham, ME).

This protocol describes the analysis of DNA content of a cell population to determine the proportion of cells that are in Go/G1, S, or G2/M phase of the cell cycle.

1. Plate cells in 60-mm dishes and grow overnight in normal medium at the appropriate incubation conditions.

2. Serum starve the cells with medium containing 0.1% BSA overnight.

3. Treat the cells with or without growth factors or with other stimuli for 12–18 h at the appropriate incubation conditions for the cell type used (*see* **Note 10**).

4. Aspirate the medium and trypsinize each dish with prewarmed 1 mL trypsin/EDTA for at least 5 min.

5. Add 2 mL of normal medium to each dish and gently, but completely, scatter the cells and collect in 15-mL tubes.

6. Collect cells by centrifuging at 300*g* (~1500 rpm) for 5 min at room temperature in a Beckman J6-HC or in a similar centrifuge.

7. Aspirate the supernatant and resuspend cells in 3 mL of ice-cold PBS.

8. Centrifuge the cells at 300*g* for 10 min at room temperature.
9. Aspirate the supernatant and gently resuspend the cells in a minimum volume (100 μL) of ice-cold PBS.
10. Fix the cells by dropwise addition of 1 mL of cold (–20°C) 70% ethanol. Gently mix the cells with ethanol by vortexing at the lowest setting while the ethanol is added and incubate at –20°C for 1 h (*see* **Note 11**).
11. Centrifuge the cells at 300*g* for 10 min at room temperature.
12. Aspirate supernatant and wash with 3 mL of ice-cold PBS three times.
13. Resuspend the cells in 1 mL of PBS. Add 100 μg/mL of RNAse (Sigma R-5503) to the cell suspension to remove the double-stranded RNA and 5 μg/mL propidium iodide to stain DNA (*see* **Note 12**).
14. Incubate the cells at 37°C for 30 min in the dark to facilitate staining.
15. Analyze the cells using a flow cytometer.

### *3.4. Evaluation of Cellular Metabolic Activity*

In general, the overall metabolic activity of cells in a pure culture should be very consistent. Thus, the ability of cells to convert a substrate to a quantifiable product should render information that correlates with the number of cells present in each well. Additionally, these assays can be performed in multiwell tissue culture plates and easily read on an absorbance plate reader that has the appropriate filters. Although the data can stand alone and the absorbance between treatments can be compared, a standard curve can be prepared and assayed with the rest of the samples so that the amount of absorbance can be converted to cell numbers.

1. Cells are plated in multiwell tissue culture plates at about 30% confluency and cultured overnight in the appropriate culture media.
2. The cells are serum starved with media containing 0.1% BSA so as to arrest them in G0 of the cell cycle.
3. After an overnight incubation, the cells are treated according to the experimental design. Changes in cell number can only be evaluated after 18–30 h, depending on the cell type and preferably over several days.
4. Prepare the MTS reagent by adding the appropriate amount of PMS reagent to the MTS. The manufacturer recommends a mixture of PMS to MTS at a ratio of 1:20.
5. After the appropriate incubation, add 20 μL of the MTS/PMS reagent per 100 μL of culture media to the wells and continue incubating the cells for 1–4 h (*see* **Note 13**).
6. Read the plate directly at 490 nm or transfer an aliquot of the media from each well to a new 96-well plate and read the samples (*see* **Note 14**).
7. Comparisons can be made by evaluating changes in the absorbance of triplicate wells between treatments.

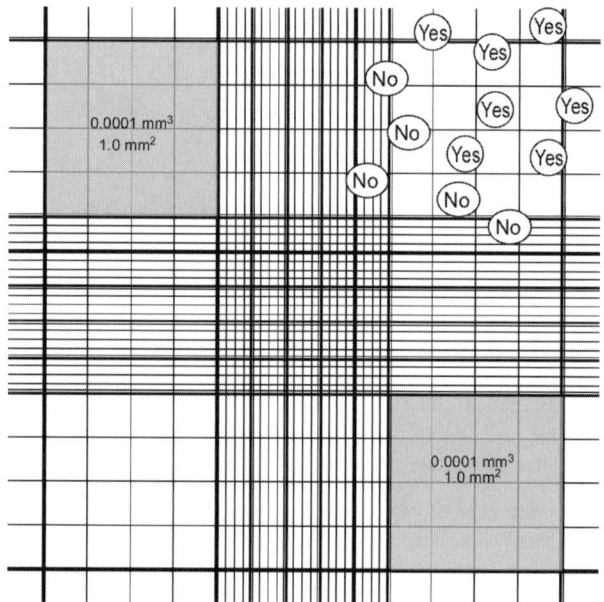

Fig. 1. The Neubauer rulings of a hemocytometer. The area of a large square is highlighted in gray. The circles containing yes and no are indicative of which cells would be counted in the upper right square.

## 4. Notes

1. Each cell type is different and requires specific growing conditions. Cells should be plated at a low enough density to allow ample room to grow and avoid contact inhibition.
2. Counting cells with a hemocytometer only measures whole cells and does not distinguish between cell cycle stages. Therefore, only cells that have completed a full cycle are counted.
3. Trypsin concentration and incubation time are cell type specific and should be determined empirically.
4. If the cell concentration is dilute, they can be concentrated by centrifugation at 300$g$ for 5 min at room temperature followed by gentle resuspension in an appropriate volume of medium such that a unicellular suspension is achieved.
5. An excessive amount of cell suspension will alter the volume in the chamber and appreciably affect the final calculated cell number (*see* **Note 7**).
6. Usually only the live cells are counted. However, a count of both live and dead cells can be used to calculate a live/dead cell ratio in the suspension.
7. Neubauer rulings in the hemocytometer cover 9 mm$^2$ (**Fig. 1**). The ruled surface is 0.10 mm below the cover glass, so that the volume of a square is 0.0001 mm$^3$. In counting the cells, care must be taken not to count any cells twice. Also, to main-

tain the precise dimensions of the area, cells that touch the lines of the outside perimeter of a square are only counted if they touch the top and right side but not the bottom or left side. Preferably, more than a total of 100 cells per chamber should be counted. In addition, it is recommended that of the nine sectors, at least the four corner and the middle sector (**Fig. 1**) be counted to correct for leveling of the microscope stage.

8. Due to the use of radioactive materials, it is a good policy to use the smallest volume possible. The deciding factor is how many cells are needed to detect [$^3$H]-thymidine incorporation. The number of cells required to achieve the optimal signal-to-noise ratio will be dependent on the cell type.

9. Radioactive materials should be disposed of according to institutional safety procedures.

10. Measurements must be performed prior to the cells completing mitosis, because those cells will be undistinguishable from cells that did not divide.

11. Alternatively, 70% methanol can also be used for fixation. The coefficient of variation (CV) of determining the cell cycle phases improves if the fixation time is extended to overnight. At this step cells can be stored at $-20°C$ for extended periods of time before further processing.

12. Propidium iodide is the most commonly used dye to quantitatively assess DNA content. However, there are a number of dyes that can be used for DNA staining including mithramycin, Hoechst 33342, 4',6-diamidino-2-phenylindole (DAPI), DRAQ5, TO-PRO-3 iodide, 7-aminoactinomycin D.

13. The incubation time will be dependent on the cell type and the number of cells present. The color change can be evaluated visually, or multiple measurements can be taken at various time points on the plate reader to assess the color development.

14. This will be dependent on the plate size that is used for the original culture and whether the plate reader that you have at your facility can read multiplate formats (i.e., 48 or 24 well).

## Acknowledgment

This work was supported by grants from the National Institutes of Health (CA105730, CA108467, and GM53271 to PJB and HL48308 and HL59679 to TBP).

## References

1. Carpenter, G. and Cohen, S. (1990) Epidermal growth factor. *J. Biol. Chem.* **265,** 7709–7712.
2. Fisher, D. A. and Lakshmanan, J. (1990) Metabolism and effects of epidermal growth factor and related growth factors in mammals. *Endocr. Rev.* **11,** 418–442.
3. Pardee, A. B. (1989) G1 events and regulation of cell proliferation. *Science* **246,** 603–608.
4. Wiepz, G. J., Houtman, J. C., Cha, D., and Bertics, P. J. (1997) Growth hormone attenuation of epidermal growth factor-induced mitogenesis. *J. Cell Physiol.* **173,** 44–53.

# 14

## Clinical Advancement of EGFR Inhibitors in Cancer Therapy

### Prakash Chinnaiyan and Paul M. Harari

#### Summary

The epidermal growth factor receptor (EGFR) represents one of the most promising molecular targets in cancer therapeutics. An array of EGFR inhibitory drugs have been developed that are progressing rapidly in oncology clinical trials. This chapter provides an overview of EGFR inhibitors and key clinical trial results that are helping to define a future role for these molecular agents in cancer treatment.

**Key Words:** Epidermal growth factor receptor inhibitors; clinical development.

## 1. Introduction

Traditional cytotoxic chemotherapy in the treatment of solid tumors is often limited by toxicity, modest efficacy, and the development of drug resistance. In particular, successful therapy options for patients with metastatic disease and tumor progression following conventional chemotherapy are limited. The last two decades of advancement in basic molecular biology have facilitated the design of molecular therapies that more specifically target tumor cells with diminished collateral damage to normal tissues (**Table 1**). One of the most promising current molecular targets for cancer therapy is the epidermal growth factor receptor (EGFR). The EGFR is overexpressed, dysregulated, or mutated in many human cancers (**Table 2**) *(1–13)*. EGFR signaling activation appears important in the growth and progression of a spectrum of malignancies and therefore holds particular appeal as a molecular target for cancer therapy *(14)* (**Fig. 1**). This chapter provides a brief overview of the clinical development of EGFR inhibitory agents in the treatment of epithelial malignancies.

## 2. EGFR Inhibitors

The cancer therapeutics field has taken several innovative approaches in an effort to develop effective EGFR inhibitors for cancer treatment. Recognizing

From: *Methods in Molecular Biology, vol. 327: Epidermal Growth Factor: Methods and Protocols*
Edited by: T. B. Patel and P. J. Bertics © Humana Press Inc., Totowa, NJ

**Table 1**
**Conventional Chemotherapy Vs Molecular Targeted Cancer Therapy**

|  | Conventional chemotherapy | Molecular targeted agents |
|---|---|---|
| Mechanism | Form covalent bonds or compete for metabolites normally incorporated into DNA/RNA<br>Target microtubules resulting in impaired meiosis | Target aberrant cell signaling pathways unique to cancer |
| Activity | Cytotoxic | Cytostatic/cytotoxic |
| Specificity | Low, generally target rapidly proliferating cells | More specific to tumors |
| Toxicity | High toxicity | Lower toxicity |

**Table 2**
**Epidermal Growth Factor Receptor (EGFR) Expression in Human Tumors**

| Tumor type | EGFR expression (%) |
|---|---|
| Head and neck | 80–100 |
| Non-small-cell lung | 40–80 |
| Colorectal | 25–77 |
| Renal | 50–90 |
| Bladder | 31–48 |
| Prostate | 40–80 |
| Esophageal | 43–89 |
| Breast | 14–91 |
| Cervical | 54–74 |
| Ovarian | 35–70 |
| Glioblastoma | 40–50 |

a central role of EGFR in cellular proliferation and differentiation, Mendelsohn and colleagues purified a series of monoclonal antibodies (MAbs) against the EGFR in the early 1980s to test these agents as inhibitors of tumor growth *(15,16)*. This simple theme has subsequently undergone tremendous expansion, particularly over the last 6–8 yr. A broad series of EGFR inhibitors have been designed, including anti-EGFR MAbs directed against the extracellular domain of the EGFR, as well as a cohort of small molecule tyrosine kinase inhibitors (TKIs) directed against the catalytic domain of the EGFR *(17,18)*. This dominant approach has been adopted by the pharmaceutical industry and has since prompted clinical development of a broad spectrum of EGFR inhibitors for testing in cancer therapy (**Table 3**).

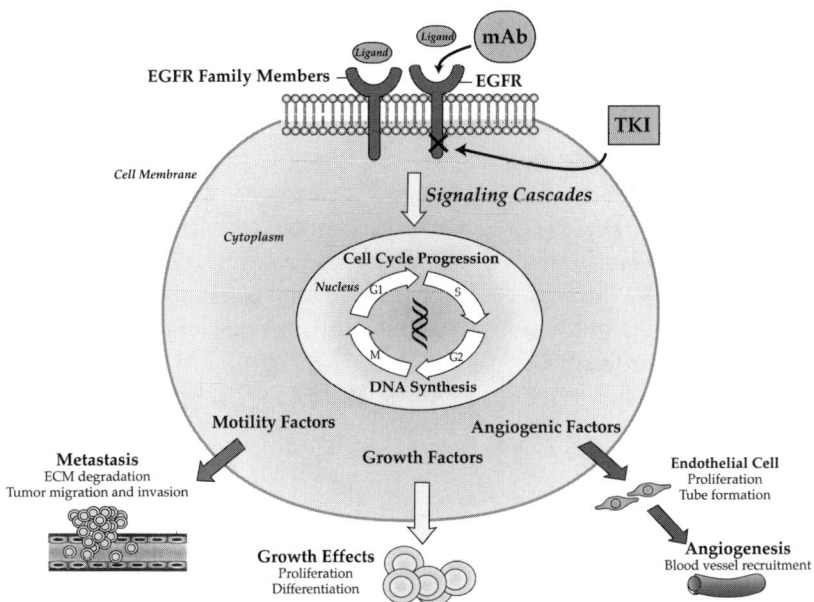

Fig. 1. Simplified schematic illustration of the epidermal growth factor receptor (EGFR) pathway highlighting potential downstream cellular and tissue effects of EGFR signaling inhibition. The action site for EGFR inhibitors is depicted for monoclonal antibodies (MAb) and tyrosine kinase inhibitors (TKI). (Adapted with permission from **ref. *14*.**)

**Table 3**
**Anti-Epidermal Growth Factor Receptor (EGFR) Agents in Clinical Development**

| Anti-EGFR agent | Description | Clinical phase |
| --- | --- | --- |
| MAb | | |
| Cetuximab (Erbitux™, C225) | Chimeric | II/III |
| ABX-EGF | Humanized | II |
| EMD 72000 | Humanized | II |
| h-R3 | Humanized | II |
| MAb 806 | Anti-EGFRvIII | I |
| TKI | | |
| Gefitinib (Iressa™, ZD1839) | Reversible | II/III |
| Erlotinib (Tarceva™, OSI-774) | Reversible | II/III |
| PKI-166 | Reversible | I |
| EKB-569 | Irreversible | I/II |
| Bispecific | | |
| Lapatinib (GW-2016) | Reversible EGFR/ErbB2 TKI | I/II |
| Canertinib (CI-1033) | Irreversible pan-Erb TKI | I/II |
| MDX-447 | EGFR/CD64 | I/II |
| ZD6474 | EGFR/VEGFR | I/II |

Currently, the MAb cetuximab (Erbitux™, ImClone) and the TKIs gefitinib (Iressa™, AstraZeneca) and erlotinib (Tarceva™, OSI-Pharma/Genentech/ Roche) represent the most mature of the EGFR inhibitors, in clinical development. However, a spectrum of EGFR inhibitors including a new generation of bispecific agents, are also progressing, thereby providing a new dimension to targeting EGFR signaling. Some examples include canertinib (CI-1033, Pfizer) and lapatinib (GW2016, GlaxoSmithKline), which target multiple ErbB family members, MDX-447 (Medarex), which binds simultaneously to EGFR and CD64 (expressed by monocytes, macrophages, and activated neutrophils) with the aim of enhancing antibody-dependent cell-mediated cytotoxicity (ADCC), and ZD6474 (AstraZeneca), which targets both EGFR and the vascular endothelial growth factor receptor (VEGFR) *(17,19,20)*.

## 3. Clinical Development

### 3.1. Phase I: Feasibility and Toxicity

Phase I trials of EGFR inhibitors demonstrate a toxicity profile quite distinct from those associated with traditional cytotoxic chemotherapy agents. The common chemotherapy toxicities of nausea, vomiting, and decreased blood counts are infrequently observed with EGFR inhibitors. Acneiform rash is the most common adverse event for the EGFR agents overall, and this appears to be a class effect relating to blockade of EGFR signaling in the skin. The acne-like rash manifests predominantly on the trunk and face, and is generally low grade and self-limiting. EGFR is constitutively expressed in epithelial tissues, and therefore skin and gastrointestinal (GI) toxicities are the most frequently reported adverse effects with the EGFR TKIs. These small molecule inhibitors readily traverse the epithelial basement membrane, thereby contributing to diarrhea. The EGFR MAbs are large molecules that do not easily cross basement membranes and thus GI toxicity is uncommon. The GI and skin toxicities typically resolve with discontinuation of the EGFR inhibitor *(21–23)*.

Pharmacokinetic studies of cetuximab report an estimated half-life of several days, making this agent suitable for once-weekly dosing. The recommended dosage of cetuximab is 400 mg/m$^2$ as an initial loading dose, with a weekly maintenance dose of 250 mg/m$^2$ *(22,24)*. Cetuximab has been tested in several phase I studies in patients with a variety of advanced solid tumors, including head and neck, lung, colorectal, prostate, kidney, breast, ovarian, pancreas, and bladder. Treatment with cetuximab in these studies was well tolerated, with the most frequent adverse events being skin toxicities, fever and chills, fatigue, and transaminase elevation. In addition, some allergic and transfusion reactions have been reported with cetuximab (2–4%), as is common to most chimeric antibodies. The induction of human antichimeric antibody and human anti-human antibody by cetuximab do not appear to be frequent or clinically significant events *(22,24)*.

The EGFR TKIs gefitinib and erlotinib have also been evaluated in phase I clinical trials. In a dose-escalating phase I trial in 64 patients with EGFR-expressing solid tumors, gefitinib was found to be well tolerated, with dose-limiting toxicity (grade 3 diarrhea) observed at doses at or above 700 mg/d. The mean terminal half-life of gefitinib was 1–2 d, compatible with a once-daily dosing schedule. The current recommended dose of gefitinib is 250 mg/d, well below the maximum tolerated dose (MTD) *(21,24,25)*. Erlotinib, which induces a similar toxicity profile as gefitinib, has been predominantly studied at doses much closer to its MTD of 150 mg/d *(23)*. The ultimate clinical impact of these different strategies in dosing of gefitinib and erlotinib is unknown, although it may be clarified in future clinical trials.

### 3.2. Phase II: Response Rates

Pivotal trials involving EGFR TKIs were largely focused on the treatment of patients with non-small-cell lung carcinoma (NSCLC). Given the lack of effective therapeutic options and the frequency of EGFR overexpression in NSCLC, evaluating the activity of EGFR inhibitors in this common and lethal disease represented a logical priority. These clinical trials provided a first glimpse of the capacity of EGFR inhibitors to treat cancer in the monotherapy setting in patients refractory to standard chemotherapy, as well as in first-line treatment when combined with conventional cytotoxic agents.

Multicenter phase II trials in patients with relapsed NSCLC following traditional chemotherapy confirmed the activity of gefitinib and erlotinib. In large-scale randomized trials (IDEAL 1 and 2 enrolling several hundred patients per trial) of gefitinib (250 mg/d vs 500 mg/d) in relapsed NSCLC patients, objective tumor response rates ranged from 8.8 to 19% *(26,27)*. In addition, approximately one-third of patients reported improvement in lung cancer symptoms assessed by the FACT-L QOL instrument *(28)*. Similar activity has been reported for erlotinib dosed at its MTD of 150 mg/d in a single-arm phase II trial in 56 evaluable patients *(29)*.

Clinical development of EGFR inhibitors in head and neck and colorectal cancer has also progressed. The phase II Bowel Oncology with Cetuximab Antibody (BOND) study represents a multicenter trial involving 329 patients to examine the potential benefit of cetuximab in irinotecan-refractory colorectal cancer patients. In this challenging population, combination treatment produced partial responses in 23% of patients and disease stabilization in an additional 33%. Median survival time was 8.6 mo, with approximately one-third of patients alive after 1 yr. Even among patients who received cetuximab alone, 11% responded to treatment. This study contributed to the Food and Drug Administration (FDA) approval of cetuximab in 2004 for use in conjunction with irinotecan in patients with irinotecan-refractory metastatic colorectal cancer or as monotherapy in patients with EGFR-expressing colorectal tumors intolerant to irinotecan-based chemotherapy *(30)*.

### 3.3. Phase III: Comparison With Standard Therapies

The promising phase II results of EGFR TKIs in NSCLC was central in stimulating subsequent phase III evaluation. In addition, early clinical experience and preclinical data demonstrating the capacity of EGFR inhibitors to enhance the activity of cytotoxic drugs prompted expectation that the addition of EGFR inhibitors to chemotherapy would further improve efficacy. Randomized phase III trials involving more than 4000 patients with advanced-stage NSCLC were performed to determine if the addition of EGFR TKIs to doublet chemotherapy would improve survival *(31–34)*. Trials were performed combining gefitinib with carboplatin/paclitaxel (INTACT 1) *(31)* and cisplatin/ gemcitabine (INTACT 2) *(32)*. These large trials completed accrual in less than 1 yr, reflecting perception that these agents might provide a therapeutic breakthrough. Unfortunately, both studies demonstrated that the addition of gefitinib (at either 250 or 500 mg/d) did not improve response rates, time to progression, or survival compared to combination chemotherapy alone *(31,32)*. These results were further corroborated by essentially identical randomized phase III trials of erlotinib in combination with carboplatin/paclitaxel (TRIBUTE trial) *(34)* or with cisplatin/gemcitabine (TALENT trial) *(33)*, which also demonstrated no global therapeutic benefit.

Caution regarding the ultimate impact of EGFR inhibitors in cancer therapy emerged following the completion of these large-scale international lung cancer trials in 2002–2003. However, optimism was renewed with the emergence of positive clinical trial results in 2004. Significant clinical emphasis had been placed on combining EGFR inhibitors with chemotherapy. However, very promising clinical data combining EGFR inhibition with radiation was developing. A phase I trial of cetuximab in combination with radiation in patients with advanced head and neck cancer showed a striking overall response rate, with 13 complete responses and 2 partial responses in 15 evaluable patients *(35)*. These results, complemented by strong preclinical data *(36,37)*, prompted larger studies investigating the combination of cetuximab in combination with radiation in head and neck squamous cell carcinoma (HNSCC). In June 2004, results of an international, randomized phase III clinical trial of 424 patients who received radiation ± cetuximab for advanced HNSCC demonstrated a near doubling of median survival for patients treated with radiation plus cetuximab, 54 mo vs 28 mo for patients treated with radiation alone. There was a statistically significant improvement ($p = 0.02$) in locoregional disease control (8% at 2 yr) and overall survival (13% at 3 yr) favoring the cetuximab arm *(38)*. In addition to providing new potential treatment options for advanced head and neck cancer patients, this pivotal trial demonstrates survival benefit using a molecularly targeted agent used as a radiation sensitizer. This finding will likely stimulate many new clinical trials for other cancer types in which radiation plays a central treatment role.

Although the addition of erlotinib to doublet chemotherapy regimens in advanced lung cancer failed to identify a survival advantage over chemotherapy alone, a recent international study does show promise in the treatment of relapsed NSCLC. This randomized phase III study examined the efficacy of monotherapy erlotinib vs placebo in advanced recurrent NSCLC patients following at least one prior chemotherapy regimen. This trial met the primary endpoint of improving overall survival from 4.7 to 6.7 mo, as well as secondary endpoints of improving time to symptomatic progression, progression-free survival, and response rate *(39)*.

### 3.4. Predictive Markers

Although the phase III trials with gefitinib in NSCLC did not improve survival, the favorable response rates and symptom improvement observed in phase II clinical trials with good safety profiles led to the approval of gefitinib in Japan in 2002 and the United States in 2003 *(40)*. The 10–15% of patients achieving an objective tumor response were often rapid in onset and durable. However, specific mechanisms of tumor response and methods to predict those patients most likely to respond have remained elusive. Retrospective analyses of patients receiving single-agent gefitinib show that responses are more frequent among non-mokers, women, and patients with bronchoalveolar carcinoma (BAC) or adenocarcinoma (ADC) with bronchoalveolar features *(41–44)*. However, no correlation between the intensity of immunohistochemical staining for EGFR and tumor response nor obvious candidate biomarker to rationally select patients for treatment has yet been confirmed.

In 2004 the first studies to identify mutations in the EGFR kinase domain that appear to confer sensitivity to gefitinib were reported *(45,46)*. It was hypothesized that patients responding to gefitinib were likely to have tumors harboring genetic alterations in specific kinases. This concept had been demonstrated by sensitivity to imatinib (Gleevec™) in chronic myeloid leukemia (CML) and gastrointestinal stromal tumor (GIST) patients harboring *bcr-abl* translocations and *c-kit* mutations, respectively, as well as sensitivity to trastuzumab (Herceptin™) in HER-2 amplified breast cancer patients. Although EGFR is widely expressed in human cancers, mutations in the receptor have not been well characterized except in brain tumors, where a significant proportion of high-grade gliomas exhibit truncation of the EGFR extracellular domain, EGFRvIII *(47)*. The most frequent reported abnormality of EGFR in human cancers prior to these studies was simply that of receptor overexpression.

The Lynch and Paez groups sequenced the EGFR gene from lung cancer specimens of patients treated (or not treated) with gefitinib as well as normal lung tissue to identify mutations *(45,46)*. These studies identified small, inframe deletions or amino acid substitutions clustered around the EGFR TK domain in 13 of 14 patients responding to gefitinib (**Fig. 2**). These studies also

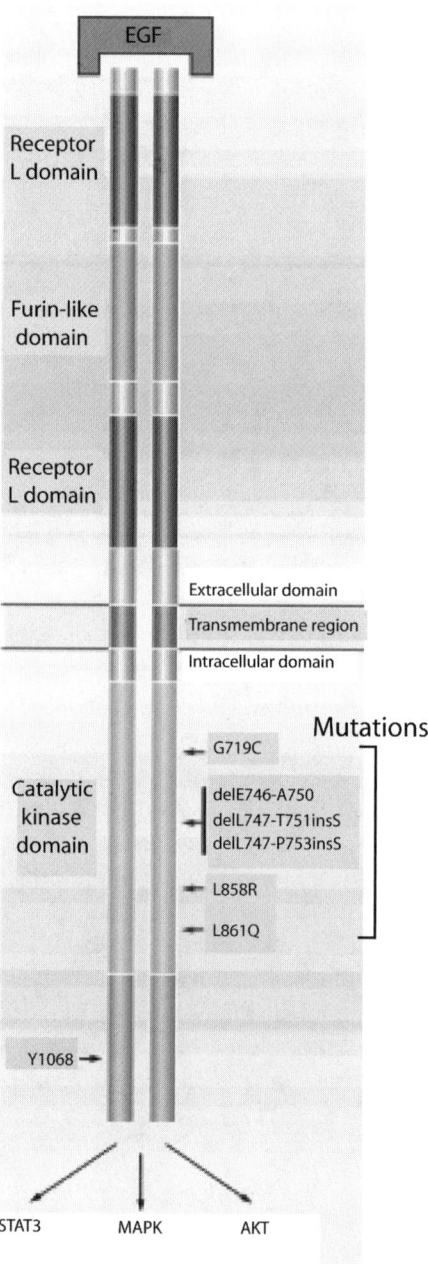

Fig. 2. Mutations identified in the epidermal growth factor receptor (EGFR) tyrosine kinase domain conferring sensitivity to gefitinib in non-small cell lung cancer patients. (Adapted with permission from **ref. 46**.)

identified that EGFR mutations were more common in ADC and BAC histologies, nonsmokers, and Japanese patients. These preliminary results complement retrospective analysis of phase II studies that describe similar characteristics of patients responding to gefitinib therapy. These studies also suggest a distinct etiology for lung cancers developing in these groups compared to squamous cell carcinomas, smokers, and non-Japanese patients. Potential biological consequences of these EGFR mutations were investigated in vitro. Cell lines harboring the mutation were 50-fold more sensitive to gefitinib than other adenocarcinoma cell lines. In addition, treatment with gefitinib completely inhibited the activation of EGFR, as well as critical downstream mediators, including extracellular signal-regulated kinase 1/2 (ERK1/2) and AKT kinase. In the absence of serum, neither wild-type nor mutant EGFR demonstrated significant autophosphorylation, although upon EGF stimulation, activation of mutant EGFRs was more robust and prolonged when compared to wild-type EGFR. These results suggest that these somatic mutations may promote cellular reliance on EGFR signaling, or an "EGFR-driven" phenotype. Further confirmation of these findings may prove valuable in predicting those patients most likely to benefit from specific EGFR inhibitor therapies.

## 4. Future Directions

Methods to better identify those tumors that rely on a particular molecular signaling pathway for their growth advantage is emerging as an important future strategy in cancer management. For EGFR, reports describing a predictive marker for sensitivity to gefitinib (involving mutations of EGFR) will stimulate further investigations. It will be important to determine if these mutations provide a similar predictive value for the anti-EGFR MAbs and other TKIs. It will also be valuable to examine for the presence of these mutations in other epithelial cancer types. There will likely emerge other mutations in the EGFR that confer either sensitivity or resistance to EGFR inhibitors. These findings will help clarify logical strategies for the ultimate use of EGFR inhibitors in cancer therapy.

Clinical strategies are also being evaluated to expand the therapeutic window of EGFR inhibitors by maximizing receptor inhibition. The rationale for this approach is supported by preclinical models demonstrating enhanced antitumor activity with maximal inhibition of EGFR signaling *(48–50)*. Examples include combining agents with complementary mechanisms of action, such as EGFR monoclonal antibodies and TKIs *(51,52)*, dual blockade of EGFR and ErbB2 *(53,54)*, and simultaneously inhibiting other oncogenic pathways including angiogenesis *(19)*, IGFR-1 signaling *(55,56)*, Hsp90 *(57)*, and histone deacetylase activity *(58)*.

Future clinical application of EGFR inhibitors may involve their capacity to enhance radiation response, as demonstrated by results of the recent phase III trial in patients with advanced HNSCC. This pivotal study is likely to stimulate further investigations to determine if this interaction between EGFR signaling and radiation response can be generalized to other tumor sites. Furthermore, investigations to better characterize molecular mechanisms underlying the interaction between EGFR signaling and radiation response will help identify patients most likely to benefit from combined EGFR/radiation treatments.

## 5. Conclusion

Although it has been more than 20 yr since the concept of targeting the EGFR in cancer therapy was initially proposed, the potential clinical value of EGFR signaling inhibitors in cancer therapy is now coming into sharper focus. Recent reports that confirm the capacity of EGFR inhibitors to improve overall survival in cancer patients will stimulate even broader interest in this evolving field. It appears likely that EGFR inhibitors (and other rationally designed molecular growth inhibitors) will play a meaningful role in cancer therapy in the years to come.

## References

1.  Rubin Grandis, J., Melhem, M. F., Barnes, E. L., and Tweardy, D. J. (1996) Quantitative immunohistochemical analysis of transforming growth factor-alpha and epidermal growth factor receptor in patients with squamous cell carcinoma of the head and neck. *Cancer* **78(6)**, 1284–1292.
2.  Salomon, D. S., Brandt, R., Ciardiello, F., and Normanno, N. (1995) Epidermal growth factor-related peptides and their receptors in human malignancies. *Crit. Rev. Oncol. Hematol.* **19(3)**, 183–232.
3.  Stumm, G., Eberwein, S., Rostock-Wolf, S., et al. (1996) Concomitant overexpression of the EGFR and erbB-2 genes in renal cell carcinoma (RCC) is correlated with dedifferentiation and metastasis. *Int. J. Cancer* **69(1)**, 17–22.
4.  Rieske, P., Kordek, R., Bartkowiak, J., Debiec-Rychter, M., Biernat, W., and Liberski, P. P. (1998) A comparative study of epidermal growth factor receptor (EGFR) and MDM2 gene amplification and protein immunoreactivity in human glioblastomas. *Pol. J. Pathol.* **49(3)**, 145–149.
5.  Itakura, Y., Sasano, H., Shiga, C., et al. (1994) Epidermal growth factor receptor overexpression in esophageal carcinoma. An immunohistochemical study correlated with clinicopathologic findings and DNA amplification. *Cancer* **74(3)**, 795–804.
6.  Fontanini, G., De Laurentiis, M., Vignati, S., et al. (1998) Evaluation of epidermal growth factor-related growth factors and receptors and of neoangiogenesis in completely resected stage I-IIIA non-small-cell lung cancer: amphiregulin and microvessel count are independent prognostic indicators of survival. *Clin. Cancer Res.* **4(1)**, 241–249.

7. Glynne-Jones, E., Goddard, L., and Harper, M. E. (1996) Comparative analysis of mRNA and protein expression for epidermal growth factor receptor and ligands relative to the proliferative index in human prostate tissue. *Hum. Pathol.* **27(7),** 688–694.

8. Ngan, H. Y., Cheung, A. N., Liu, S. S., Cheng, D. K., Ng, T. Y., and Wong, L. C. (2001) Abnormal expression of epidermal growth factor receptor and c-erbB2 in squamous cell carcinoma of the cervix: correlation with human papillomavirus and prognosis. *Tumour Biol.* **22(3),** 176–183.

9. Chow, N. H., Chan, S. H., Tzai, T. S., Ho, C. L., and Liu, H. S. (2001) Expression profiles of ErbB family receptors and prognosis in primary transitional cell carcinoma of the urinary bladder. *Clin. Cancer Res.* **7(7),** 1957–1962.

10. Kersemaekers, A. M., Fleuren, G. J., Kenter, G. G., et al. (1999) Oncogene alterations in carcinomas of the uterine cervix: overexpression of the epidermal growth factor receptor is associated with poor prognosis. *Clin. Cancer Res.* **5(3),** 577–586.

11. Bartlett, J. M., Langdon, S. P., Simpson, B. J., et al. (1996) The prognostic value of epidermal growth factor receptor mRNA expression in primary ovarian cancer. *Br. J. Cancer* **73,** 301–306.

12. Fischer-Colbrie, J., Witt, A., Heinzl, H., et al. (1997) EGFR and steroid receptors in ovarian carcinoma: comparison with prognostic parameters and outcome of patients. *Anticancer Res.* **17(1B),** 613–619.

13. Klijn, J. G., Berns, P. M., Schmitz, P. I., and Foekens, J. A. (1992) The clinical significance of epidermal growth factor receptor (EGF-R) in human breast cancer: a review on 5232 patients. *Endocr. Rev.* **13(1),** 3–17.

14. Harari, P. M. and Huang, S. M. (2002) Epidermal growth factor receptor modulation of radiation response: preclinical and clinical development. *Semin. Radiat. Oncol.* **12(3 Suppl 2),** 21–26.

15. Kawamoto, T., Sato, J. D., Le, A., Polikoff, J., Sato, G. H., and Mendelsohn, J. (1983) Growth stimulation of A431 cells by epidermal growth factor: identification of high-affinity receptors for epidermal growth factor by an anti-receptor monoclonal antibody. *Proc. Natl. Acad. Sci. USA* **80(5),** 1337–1341.

16. Mendelsohn, J. (2003) Antibody-mediated EGF receptor blockade as an anticancer therapy: from the laboratory to the clinic. *Cancer Immunol. Immunother.* **52(5),** 342–346. Epub 2003 Mar 14.

17. Arteaga, C. (2003) Targeting HER1/EGFR: a molecular approach to cancer therapy. *Semin. Oncol.* **30(3 Suppl 7),** 3–14.

18. Mendelsohn, J. and Baselga, J. (2003) Status of epidermal growth factor receptor antagonists in the biology and treatment of cancer. *J. Clin. Oncol.* **21(14),** 2787–2799.

19. Ciardiello, F., Caputo, R., Damiano, V., et al. (2003) Antitumor effects of ZD6474, a small molecule vascular endothelial growth factor receptor tyrosine kinase inhibitor, with additional activity against epidermal growth factor receptor tyrosine kinase. *Clin. Cancer Res.* **9(4),** 1546–1556.

20. Raymond, E., Faivre, S., and Armand, J. P. (2000) Epidermal growth factor receptor tyrosine kinase as a target for anticancer therapy. *Drugs* **60(Suppl 1),** 15–23; discussion 41–42.

21. Baselga, J., Rischin, D., Ranson, M., et al. (2002) Phase I safety, pharmacokinetic, and pharmacodynamic trial of ZD1839, a selective oral epidermal growth factor receptor tyrosine kinase inhibitor, in patients with five selected solid tumor types. *J. Clin. Oncol.* **20(21),** 4292–4302.

22. Baselga, J., Pfister, D., Cooper, M. R., et al. (2000) Phase I studies of anti-epidermal growth factor receptor chimeric antibody C225 alone and in combination with cisplatin. *J. Clin. Oncol.* **18(4),** 904–914.

23. Hidalgo, M., Siu, L. L., Nemunaitis, J., et al. (2001) Phase I and pharmacologic study of OSI-774, an epidermal growth factor receptor tyrosine kinase inhibitor, in patients with advanced solid malignancies. *J. Clin. Oncol.* **19(13),** 3267–3279.

24. Thomas, S. M. and Grandis, J. R. (2004) Pharmacokinetic and pharmacodynamic properties of EGFR inhibitors under clinical investigation. *Cancer Treat. Rev.* **30(3),** 255–268.

25. Ranson, M., Hammond, L. A., Ferry, D., et al. (2002) ZD1839, a selective oral epidermal growth factor receptor-tyrosine kinase inhibitor, is well tolerated and active in patients with solid, malignant tumors: results of a phase I trial. *J. Clin. Oncol.* **20(9),** 2240–2250.

26. Kris, M. G., Natale, R. B., Herbst, R. S., et al. (2002) A phase II trial of ZD1839 ('Iressa') in advanced non-small cell lung cancer (NSCLC) patients who had failed platinum- and docetaxel-based regimens (IDEAL 2). *Proc. Am. Soc. Clin. Oncol.* **21,** A#1166.

27. Fukuoka, M., Yano, S., Giaccone, G., et al. (2002) Final results from a phase II trial of ZD1839 ('Iressa') for patients with advanced non-small-cell lung cancer (IDEAL 1). *Proc. Am. Soc. Clin. Oncol.* **21,** A#1188

28. Natale, R. B. and Zaretsky, S. L. (2002) ZD1839 (Iressa): what's in it for the patient? *Oncologist* **7(Suppl 4),** 25–30.

29. Perez-Soler, R., Chachoua, A., Huberman, M., et al. (2001) A phase II trial of the epidermla growth factor receptor (EGFR) tyrosine kinase inhibitor OSI-774 following platinum based chemotherapy in patients with advanced EGFR expressing non-small cell lung cancer (NSCLC). *Proc. Am. Soc. Clin. Oncol.* **20,** 310.

30. Cunningham, D., Humblet, Y., Siena, S., et al. (2003) Cetuximab (C225) alone or in combination with irinotecan (CPT-11) in patients with epidermal growth factor receptor (EGFR)-positive, irinotecan-refractory metastatic colorectal cancer (MCRC). *Proc. Am. Soc. Clin. Oncol.* **22,** 252.

31. Giaccone, G., Herbst, R. S., Manegold, C., et al. (2004) Gefitinib in combination with gemcitabine and cisplatin in advanced non-small-cell lung cancer: a phase III trial—INTACT 1. *J. Clin. Oncol.* **22(5),** 777–784.

32. Herbst, R. S., Giaccone, G., Schiller, J. H., et al. (2004) Gefitinib in combination with paclitaxel and carboplatin in advanced non-small-cell lung cancer: a phase III trial—INTACT 2. *J. Clin. Oncol.* **22(5),** 785–794.

33. Gatzemeier, U., Pluzanska, A., Szczesna, A., et al. (2004) Results of a phase III trial of erlotinib (OSI-774) combined with cisplatin and gemcitabine (GC) chemotherapy in advanced non-small cell lung cancer (NSCLC). *Proc. Am. Soc. Clin. Oncol.* **23.**

34. Herbst, R. S., Prager, D., Hermann, R., et al. (2004) TRIBUTE—A phase III trial of erlotinib HCl (OSI-774) combined with carboplatin and paclitaxel (CP) chemotherapy in advanced non-small cell lung cancer (NSCLC). *Proc. Am. Soc. Clin. Oncol.* **23**.

35. Robert, F., Ezekiel, M. P., Spencer, S. A., et al. (2001) Phase I study of anti—epidermal growth factor receptor antibody cetuximab in combination with radiation therapy in patients with advanced head and neck cancer. *J. Clin. Oncol.* **19(13)**, 3234–3243.

36. Huang, S. M., Bock, J. M., and Harari, P. M. (1999) Epidermal growth factor receptor blockade with C225 modulates proliferation, apoptosis, and radiosensitivity in squamous cell carcinomas of the head and neck. *Cancer Res.* **59(8)**, 1935–1940.

37. Milas, L., Mason, K., Hunter, N., et al. (2000) In vivo enhancement of tumor radioresponse by C225 antiepidermal growth factor receptor antibody. *Clin. Cancer Res.* **6(2)**, 701–708.

38. Bonner, J. A., Giralt, J., Harari, P. M., et al. (2004) Cetuximab prolongs survival in patients with locregionally advanced squamous cell carcinoma of head and neck: A phase III study of high dose radiation therapy with or without cetuximab. *Proc. Am. Soc. Clin. Oncol.* **23**.

39. Shepherd, F. A., Pereira, J., Ciuleanu, T. E., et al. (2004) A randomized placebo-controlled trial of erlotinib inpatients with advanced non-small cell lung cancer (NSCLC) following failure of 1st line or 2nd line chemotherapy. A National Cancer Institute of Canada Clinical Trials Group (NCIC CTG) Trial. *Proc. Am. Soc. Clin. Oncol.* **23**.

40. Cohen, M. H., Williams, G. A., Sridhara, R., et al. (2004) United States Food and Drug Administration Drug Approval summary: Gefitinib (ZD1839; Iressa) tablets. *Clin. Cancer Res.* **10(4)**, 1212–1218.

41. Miller, V. A., Kris, M. G., Shah, N., et al. (2004) Bronchioloalveolar pathologic subtype and smoking history predict sensitivity to gefitinib in advanced non-small-cell lung cancer. *J. Clin. Oncol.* **22(6)**, 1103–1109.

42. Kris, M. G., Natale, R. B., Herbst, R. S., et al. (2003) Efficacy of gefitinib, an inhibitor of the epidermal growth factor receptor tyrosine kinase, in symptomatic patients with non-small cell lung cancer: a randomized trial. *JAMA* **290(16)**, 2149–2158.

43. Janmaat, M. L., Kruyt, F. A., Rodriguez, J. A., and Giaccone, G. (2003) Response to epidermal growth factor receptor inhibitors in non-small cell lung Cancer cells: limited antiproliferative effects and absence of apoptosis associated with persistent activity of extracellular signal-regulated kinase or Akt kinase pathways. *Clin. Cancer Res.* **9**, 2316–2326.

44. Fukuoka, M., Yano, S., Giaccone, G., et al. (2003) Multi-institutional randomized phase II trial of gefitinib for previously treated patients with advanced non-small-cell lung cancer. *J. Clin. Oncol.* **21(12)**, 2237–2246. Epub 2003 May 14.

45. Paez, J. G., Janne, P. A., Lee, J. C., et al. (2004) EGFR mutations in lung cancer: correlation with clinical response to gefitinib therapy. *Science* **304(5676)**, 1497–1500. Epub 2004 Apr 29.

46. Lynch, T. J., Bell, D. W., Sordella, R., et al. (2004) Activating mutations in the epidermal growth factor receptor underlying responsiveness of non-small-cell lung cancer to gefitinib. *N. Engl. J. Med.* **350(21),** 2129–2139. Epub 2004 Apr 29.

47. Ekstrand, A. J., James, C. D., Cavenee, W. K., Seliger, B., Pettersson, R. F., and Collins, V. P. (1991) Genes for epidermal growth factor receptor, transforming growth factor alpha, and epidermal growth factor and their expression in human gliomas in vivo. *Cancer Res.* **51(8),** 2164–2172.

48. Akita, R. W., Dugger, D., Philips, G., et al. (2002) Effects of the EGFR inhibitor TARCEVA on activated ErbB2. *Proc. Am. Assoc. Cancer Res.* **43,** 1003.

49. Pollack, V. A., Savage, D. M., Baker, D. A., et al. (1999) Inhibition of epidermal growth factor receptor-associated tyrosine phosphorylation in human carcinomas with CP-358,774: dynamics of receptor inhibition in situ and antitumor effects in athymic mice. *J. Pharmacol. Exp. Ther.* **291,** 739–748.

50. Hidalgo, M. (2003) Erlotinib: preclinical investigations. *Oncology* (Huntingt) **17(11 Suppl 12),** 11–16.

51. Huang, S. M., Armstrong, E., Chinnaiyan, P., and Harari, P. M. (2003) Dual agent molecular targeting of the epidermal growth factor receptor: Combining anti-HER1/EGFR monoclonal antibody with tyrosine kinase inhibitor. *Proc. Am. Assoc. Cancer Res.* **44.**

52. Matar, P., Rojo, F., Guzman, M., Rodriguez-Viltro, I., Arribas, J., and Baselga, J. (2003) Combined anti-epidermal growth factor receptor (EGFR) treatment with a tyrosine kinase inhibitor gefitinib (ZD1839, 'Iressa') and a monoclonal antibody (IMC-C225): Evidence of synergy. *Proc. Am. Assoc. Cancer Res.* **44.**

53. Moulder, S. L. and Arteaga, C. L. (2003) A Phase I/II Trial of trastuzumab and gefitinib in patients with Metastatic Breast Cancer that overexpresses HER2/neu (ErbB-2). *Clin. Breast Cancer* **4(2),** 142–145.

54. Slichenmyer, W. J., Elliott, W. L., and Fry, D. W. (2001) CI-1033, a pan-erbB tyrosine kinase inhibitor. *Semin. Oncol.* **28,** 80–85.

55. Lu, D., Zhang, H., Ludwig, D., et al. (2004) Simultaneous blockade of both the epidermal growth factor receptor and the insulin-like growth factor receptor signaling pathways in cancer cells with a fully human recombinant bispecific antibody. *J. Biol. Chem.* **279(4),** 2856–2865. Epub 2003 Oct 23.

56. Chakravarti, A., Loeffler, J. S., and Dyson, N. J. (2002) Insulin-like growth factor receptor I mediates resistance to anti-epidermal growth factor receptor therapy in primary human glioblastoma cells through continued activation of phosphoinositide 3-kinase signaling. *Cancer Res.* **62(1),** 200–207.

57. Citri, A., Kochupurakkal, B. S., and Yarden, Y. (2004) The achilles heel of ErbB-2/HER2: regulation by the Hsp90 chaperone machine and potential for pharmacological intervention. *Cell Cycle* **3(1),** 51–60.

58. Chinnaiyan, P., Valabhaneni, G., Armstrong, E., Huang, S., and Harari, P. M. (2004) Enhancing the anti-tumor activity of ErbB blockade with histone deacetylase (HDAC) inhibition. *Proc. Am. Soc. Clin. Oncol.* **23.**

# Index

## A

Acridinium-linked immunosorbent
  assay (ALISA), soluble
  epidermal growth factor
  receptor,
interpretation, 44, 45
monoclonal antibody labeling, 42, 43
principles, 40–42
rationale, 40
reaction conditions and detection,
  43–46
ADAM metalloproteinases,
  ectodomain shedding role, *see*
    Ectodomain shedding,
    epidermal growth factor
    receptor ligands
  epidermal growth factor receptor
    transactivation studies,
    dominant negative ADAM
    construct studies, 92, 93, 95,
    96
  flow cytometry, 89, 90, 95, 96
  inhibitor studies, 89, 95
  ligand detection in cell culture
    medium, 90–92, 96
  materials, 86–89, 96
  overview, 86
  RNA interference studies, 92, 96
ALISA, *see* Acridinium-linked
  immunosorbent assay
Amphiregulin,
  ectodomain shedding, *see*
    Ectodomain shedding,
    epidermal growth factor
    receptor ligands
  epidermal growth factor receptor
    transactivation studies,

flow cytometry, 89, 90, 95, 96
inhibitor studies, 89, 95
ligand detection in cell culture
  medium, 90–92, 96
materials, 86–89, 96
overview, 86
RNA interference studies, 92, 96

## B, C

Boyden chamber assay, epithelial cell
  migration, 148, 155–157
Cancer,
  CXCR4 expression, 139, 140
  epidermal growth factor receptor
    inhibitor therapy, *see* Clinical
    trials, epidermal growth factor
    receptor inhibitors
  epidermal growth factor receptor
    overexpression and mutations,
    12, 13, 26, 39, 40, 189, 190
Carbachol, epidermal growth factor
  receptor transactivation, 9
Cbl, *see* Ubiquitination
Cell migration, *see also* Epithelial cell
  migration,
  contractile force generation,
    assays, 162, 163, 165, 169–171,
    174
    overview, 162
  de-adhesion assays, 163, 165,
    171–175
  lamellipod extension assay, 161,
    164, 165, 169
  motility assays,
    approaches, 160, 161
    materials, 163, 164
    parameters in measurement, 160